Environmental Risk Assessment

Edited by

Diana Mariana Cocârță
Department of Energy Production and Use, Faculty of Energy Engineering, University POLITEHNICA of Bucharest, Romania

Environmental Risk Assessment

Editor: Diana Mariana Cocârță

ISBN (Online): 978-981-5179-39-2

ISBN (Print): 978-981-5179-40-8

ISBN (Paperback): 978-981-5179-41-5

need for a court order if at any point you breach any terms of this License Agreement. In no event will any delay or failure by Bentham Science Publishers in enforcing your compliance with this License Agreement constitute a waiver of any of its rights.

3. You acknowledge that you have read this License Agreement, and agree to be bound by its terms and conditions. To the extent that any other terms and conditions presented on any website of Bentham Science Publishers conflict with, or are inconsistent with, the terms and conditions set out in this License Agreement, you acknowledge that the terms and conditions set out in this License Agreement shall prevail.

Bentham Science Publishers Pte. Ltd.
80 Robinson Road #02-00
Singapore 068898
Singapore
Email: subscriptions@benthamscience.net

BENTHAM SCIENCE

CONTENTS

This work of developing the eBook content was supported by the Erasmus+ Programme SafeEngine project, contract no 2020-1-RO01-KA203-080085.

The European Commission's support for this publication does not constitute an endorsement of the contents, which reflects the views of the authors, and the National Agency and Commission cannot be held responsible for any use which may be made of the information contained therein.

FOREWORD

Environmental pollution is a major issue affecting both industrialized and developing countries. According to some recent studies, there are more than five million contaminated sites worldwide. In Europe, there are around 2.5 million potentially contaminated sites and over 340,000 contaminated sites; the management of these sites involves costs of € 6.5 billion per year, covered by private companies, according to the "polluter pays" principle, but also by public funds.

Sources of contamination can be natural, for example, volcanic emissions and eruptions, continental dust transport, and metal-rich rock weathering. The main sources, however, are of anthropogenic origin and include industrial processes and mining, poor waste management, unsustainable farming practices, accidents such as chemical spills, and even armed conflicts.

Those activities generate wastes and emissions that contain toxic substances and, if not properly managed, cause the diffusion and accumulation of pollutants in the soil, subsoil, and groundwater. Their contamination generates significant negative impacts on the ecosystem and human health, causing loss of biodiversity and disabling diseases, which can even lead to the death of people. This also compromises the soil's ability to provide ecosystem services, including the production of safe food. More generally, soil and subsoil pollution hinders the achievement of many of the United Nations Sustainable Development Goals (SDGs), including those related to SDG 1 (poverty elimination), SDG 2 (zero hunger), and SDG 3 (good health and well-being). Soil pollution strikes the most vulnerable people, especially children and women (SDG 5) and the supply of safe drinking water (SDG 6), which is threatened by the leaching of contaminants into groundwater and runoff. Moreover, CO_2 and N_2O emissions from unproperly managed soil cause climate change (SDG 13), and soil pollution contributes to land degradation and loss of terrestrial (SDG 15) and aquatic (SDG 14) biodiversity, and reduces the security and resilience of cities (SDG 11).

The presence of one or more contaminants in the soil and/or the groundwater does not in itself pose a hazard. The state of contamination can be assessed through three different criteria: the comparison with natural background concentrations of pollutants, the comparison with threshold concentrations, the human health risk assessment (HRA) and the ecological risk assessment.

HRA is a method for assessing the possible harm caused by contaminant emissions that affect human health. Its origin dates to the 1950s, but the first concrete applications took place in the United States in the 1980s, after the National Academy of Science published "Risk Assessment in the Federal Government: Managing the Process" in 1983. Later, in 1992, the US Environmental Protection Agency's "Framework for Ecological Risk Assessment" introduced a simple and flexible structure for conducting and evaluating ecological risk assessment.

Much has come since then, but there is still much to be done. In the European Union, for instance, there is currently neither a univocal definition of "contaminated site" nor a Directive concerning the remediation of contaminated sites. Many Member States have their own legislation and have adopted different definitions, which are not homogeneous. The health and environmental risk assessment often has a key role as it represents a fundamental decision-making tool not only in the assessment of contamination but also in the selection and implementation of the remediation strategies, which can include containment works, remedial actions or monitored natural attenuation approaches.

This book, which depicts a complete and up-to-date picture of the environmental risk assessment, represents a very useful tool for technicians and decision-makers working in the environmental field, who will be guided through the methodologies and procedures that can be used to implement the risk-based approach for contaminated soil management, air and drinking water quality protection, and waste management.

Mentore Vaccari
University of Brescia, Italy

PREFACE

Environmental pollution has been a topic of growing interest all over the world, in both developed and developing countries. At the global level, decision makers are constantly trying to identify sustainable solutions for environmental pollution issues. Multiple international agreements have been adopted to set out a global framework to avoid the dangerous effects of environmental pollution. All these actions also aim to strengthen countries' ability to deal with the impact of environmental pollution and support them in their efforts to mitigate it.

With a rapid increase in population, the demand for energy, food production, machine development, and increasing trends of urbanization has resulted in serious soil, water, and air pollution that affects the surrounding environment and includes human health. According to the World Health Organization (WHO), 24% of all estimated global mortalities are linked to environmental pollution. *Sustainable development* has been defined in many ways, but the most frequently quoted definition is from *Our Common Future*: "*Sustainable development* is a development that meets the needs of the present, without compromising the ability of future generations to meet their own needs". In this context, according to the World Health Organization, every day, approximately 93% of the world's children under the age of 15 years (1.8 billion children) breathe air that is so polluted; it puts their health and development at serious risk. The deaths of 297,000 children aged under 5 years could be avoided each year if risk factors like unsafe drinking water, sanitation, and hand hygiene are addressed.

On the other hand, environmental pollution has led to serious disruptions in natural systems: *e.g.*, snow and ice are melting, hydrological and biological systems are changing, and negative pollution effects are not stopping here. The consequences of environmental pollution for biodiversity and ecosystem conservation have also been observed in the degradation of the benefits that natural ecosystems provide for society, named ecosystem services. Examples of ecosystem services include products such as food, fibres, fuels and water; regulation of air quality and soil fertility; control of floods, soil erosion, crop pollination, and disease outbreaks; and non-material benefits such as recreational, cultural and spiritual benefits in natural areas.

The proposed book has the main aim of a broad vision of the main environmental systems: soil, water, and air. The chapters are focused on a risk-based approach to the environment and a deep dive into risk management implementation, risk considering contaminated sites, air quality, safe drinking water, and the importance of risk analysis in waste management, followed by good practices considering environmental hazards and tools in assessing risks on human's health.

In concordance with the sustainable development definition, the environment must be protected and sustainably managed. This responsibility is ours, together, we should create an educated and correctly informed society regarding environmental protection. Aware of the multiple benefits of a clean environment on human health, our actions as individuals and societies should be only in the environmental protection direction. Stefania Giannini (UNESCO Assistant Director-General for Education) said that "*through education, we could create a peaceful and sustainable world for the survival and prosperity of current and future generations*". In this context, the proposed book represents a guideline for students that study in the environmental engineering fields. The book aims to enable learners to develop knowledge and awareness about environmental risk management and take action to transform society into a more sustainable one. Developing an educated and correctly informed society is a top priority because it is the foundation on which we build peace and drive sustainable development.

Diana Mariana COCÂRȚĂ
Department of Energy Production and Use
Faculty of Energy Engineering
University POLITEHNICA of Bucharest
Romania

List of Contributors

A.M. Velcea	University POLITEHNICA of Bucharest, Faculty of Energy Engineering, Splaiul Independentei 313, RO-060042 Bucharest, Romania
C. Streche	University POLITEHNICA of Bucharest, Faculty of Energy Engineering, Splaiul Independentei 313, RO-060042 Bucharest, Romania
C. Stan	Department of Energy Production and Use, Faculty of Energy Engineering, University POLITEHNICA of Bucharest, Bucharest, Romania
Diana Mariana Cocârță	University POLITEHNICA of Bucharest, Faculty of Energy Engineering, Splaiul Independentei 313, RO-060042 Bucharest, Romania
F.P. Martín Jiménez	Department of Chemical Engineering, University of Malaga, Faculty of Science, Malaga, Spain
Lăcrămioara D. Robescu	Department of Hydraulics, Hydraulic Machinery and Environmental Engineering, Faculty of Energy Engineering, University Politehnica of Bucharest, Bucharest, Romania
M.C. López-Escalante	Department of Chemical Engineering, University of Malaga, Faculty of Science, Malaga, Spain
Marius D. Bontoș	Department of Hydraulics, Hydraulic Machinery and Environmental Engineering, University POLITEHNICA of Bucharest, Faculty of Energy Engineering,, Bucharest, Romania
R. Lupu	University POLITEHNICA of Bucharest, Faculty of Energy Engineering, Splaiul Independentei 313, RO-060042 Bucharest, Romania

CHAPTER 1

Environmental Pollution and Health

Diana Mariana Cocârță[1,*] and **A.M. Velcea**[1]

[1] *University POLITEHNICA of Bucharest, Faculty of Energy Engineering, Splaiul Independentei 313, RO-060042 Bucharest, Romania*

Abstract: Both developed and developing nations around the world are becoming increasingly interested in environmental pollution and impact human health. Different factors contribute to environmental pollution, including an increase in population, resulting in demand for energy, which causes toxic pollutants that are released into the *air we breathe*, on the *soil where we grow food*, and in the *water we drink*. These contaminants may be harmful to both the environment and human health.

The influence of environmental pollution on human health and well-being is discussed in detail in the current chapter. There are examples of various environmental problems related to soil, air, and water pollution, as well as evidence of human exposure pathways and the health effects of different environmental pollutants. Specific chemical contaminants present in soil, air and water are also evidenced. So, this chapter introduces the reader to a world where environmental health is synonymous with human health and where how each of us as individuals treats the environment directly affects our well-being.

Keywords: Air pollution, Contaminants, Environmental Risk Assessment, Environmental pollution, Ecological Risk Assessment, Human exposure, Human Health Risk Assessment, Risk-based approach, Soil pollution, Water pollution.

INTRODUCTION

The Environmental or Ecological Risk Assessment study is mainly focused on understanding the potential negative effects of human activities on the ecosystem (plants, animals, lakes, and seas).

This book is focused on key elements of Environmental Risk Assessment, how to manage or to perform such study, in the context of Air, Water or Soil pollution, as the primary source of investigation. The aim of the chapters is to promote a structured approach to Environmental Risk Assessment (ERA), provide high-quality information that is consistent with good practices and, most importantly,

---
* **Corresponding author Diana Mariana Cocârță:** University POLITEHNICA of Bucharest, Faculty of Energy Engineering, Splaiul Independentei 313, RO-060042 Bucharest, Romania; E-mail: dianacocarta13@yahoo.com

keep on alert engineering students and decision-makers about the environmental problems for controlling and applying corrective measures to minimise risk and/or to avoid risk occurrence.

ENVIRONMENTAL POLLUTION

The environment is composed of lithosphere (rocks and soil), hydrosphere (water), atmosphere (air) and biosphere (living component of the environment). Environmental Pollution is described as an excessive amount of harmful chemicals in the environment (water, air, and soil), making it dangerous for life. All sources of contaminants, as an initial step, are discharged in one of the environmental components. The contaminants further go through physical and chemical changes, which are lastly incorporated in the medium [1]. For instance, once the pollutants are emitted into the atmosphere, a conversion principle is applied: "*Matter cannot be destroyed; it is merely converted from one form to another*" [2], known as the second law of thermodynamics. In other words, the contaminants that reach the environment are dispersed based on their properties, medium characteristics, and others, and further can be converted (or not) into another type of substance. This type of conversion is applied to the substances/materials which can be replaced or renewed, and these substances/materials, once in the environmental media, easily are assimilated and do not interfere with the well-being of the environment [1].

To understand the meaning of the pollution, it is important to define the characteristics of the pollutants present in the environment and what effects does it have. Contaminants can occur from diverse sources, natural or man-made. Natural pollution results from different sources such as wildfires, volcanic activity, or seismic activities. In the case of anthropic pollution, this derives from human activity. Examples in this regard are: untreated industrial and municipal wastewater discharge, burning of the fossil fuel, which leads to the atmospheric increase of CO_2 and other greenhouse gases, increasing the global warming and climate change effects at the global level, uncontrolled dumping of waste, excessive applications of chemical fertilizers and pesticides on agricultural soils, or accidental spills of toxic organic substances in the soil (petroleum products, chlorinated solvents). These substances are able to move from one environment system (soil, water, air) towards another through migration processes like: leaching, volatilization, photo-decomposition, runoff, wet and dry deposition, *etc.*, [1]. Common toxic substances found in the environment are illustrated in Table **1.1**:

Table 1.1. Most common chemical contaminants present in the environment [3].

Chemical Classification	Frequency of Occurrence
Gaseous contaminants CO_x (CO, CO_2), NO_x (NO, NO_2, N_2O), SO_x (SO, SO_2, SO_3), NH_3, VOC_s	Very frequent
Gasoline, fuel oil	Very frequent
Alcohols, ketones, esters	Common
Chlorinated organics	Very frequent
Polychlorinated biphenyls (PCB_s)	Infrequent
Nitroaromatics	Common
Metals (Cd, Cr, Cu, Hg, Ni, Pb, Zn)	Common
Nitrate	Common
Phosphate	Common
Ethers	Common

ENVIRONMENTAL POLLUTION AND IMPACT

Air Pollution

Air pollution is one of the biggest issues all around the world in both developed and developing countries, and is mainly caused by heavy traffics, rapid development of the economy, industrialization, exploitation of natural resources and so on. The rapid growth of population and demand for food, energy and materials have driven the emissions of various toxic compounds into the air, impacting human and ecosystem health [4]. Based on the report from World Health Organization, 384 million people suffer from chronic obstructive pulmonary disease, and around 3 million death cases result annually, along with other respiratory health issues caused by air pollution. This issue is leading to the third cause of death worldwide [5].

According to World Health Organization, the cities shall be evaluated considering the air quality based on the average level of particulate matter ($PM_{2.5}$) in the air. Fine particles ($PM_{2.5}$) pose the greatest health risk because these particles have a very small size (particle diameter<2.5 μm) and can get deep into lungs, and some may even get into the bloodstream. Health effects may include cardiovascular effects, such as cardiac arrhythmias and heart attacks, and respiratory effects, such as asthma attacks and bronchitis. Exposure to particle pollution affects especially the population with pre-existing heart or lung diseases, older people, and children. According to Statista Company, the most polluted 10 countries in the world are presented in Fig. (**1.1**); for 2020 (light blue color) and 2021 (dark blue color), based on the levels of particulate matter ($PM_{2.5}$) present in the air (in $\mu g/m^3$):

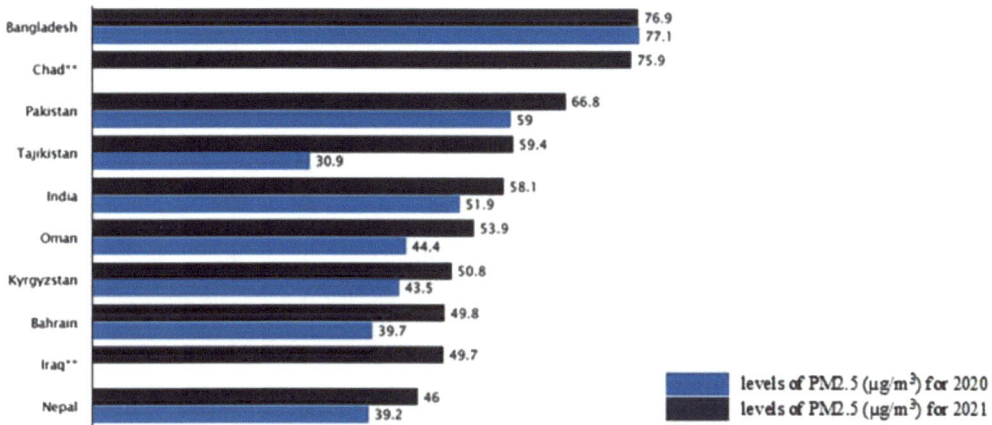

Fig. (1.1). The most polluted countries in the world in 2020 and 2021, according to Statista Company (https://www.statista.com/statistics/1135356/most-polluted-countries-in-the-world/).

In 2021, WHO elaborated Guidelines for air quality emissions in accordance with human health protection and recommended a maximum level of 5 $\mu g/m^3$ of fine particle ($PM_{2.5}$) in the air for long-term exposure of humans. According to the Directive 2008/50/EC for clean air, European Union has set a level of 25 $\mu g/m^3$ of PM_{10}, however, the Directive is now under revision to align more closely to WHO Standards. Currently, European Environmental Agency has proposed a wide and comprehensive air quality monitoring for particulate matter. Data achieved from the monitoring stations of around 340 cities in the European Union can be visualized in Fig. (**1.2**).

Fig. (**1.2**) illustrates that Spain has predominantly good to moderate air quality. Moving to Italy, the northern part is characterized as moderate to poor (some regions have an air quality over 25 $\mu g/m^3$ imposed by the EU Directive for PM_{10} and $PM_{2.5}$ [6]). In Romania, air quality monitoring can be visualized using different online free websites financed by the ONG's and Government. Examples of websites illustrating real-time data on air quality from monitoring stations across Romania are listed below:

1. https://aqicn.org/map/romania/

2. https://www.calitateaer.ro/public/home-page/?__locale=ro

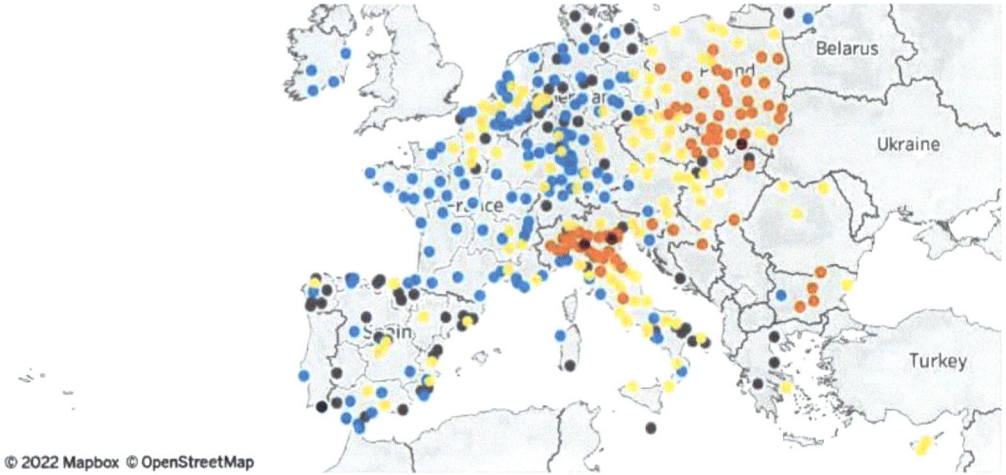

(a)

PM2.5 annual mean concentration,
µg/m3

0 - 5	good	■
5 - 10	fair	■
10 - 15	moderate	■
15 - 25	poor	■
> 25	very poor	■
no data	-	■

(b)

Fig. (1.2). Air quality in different cities across the European Union using the particulate matter monitoring stations results (**a**), the level of PM associated with the color (**b**) (source: https://www.eea.europa.eu/themes/air/urban-air-quality/european-city-air-quality-viewer).

The most well-known substances contributing to air pollution and with important evidence regardings the negative consequences on public health are nitrogen dioxide (NO_2), sulfur dioxide (SO_2), carbon monoxide (CO), particulate matter

(PM), volatile organic compounds (VOCs) and ozone (O_3). Particles and vapours emitted in the atmosphere can occur year-round, which is considered a big issue for concentrations found in many major cities throughout Europe and in other parts of the world. Some particles from the atmosphere can remain for longer periods, days, or weeks. Moreover, particles, once emitted into the atmosphere, can travel hundreds or thousands of kilometers and affect the air quality of the zones far from the original source [7]. There are different circumstances in which high levels of particle pollution can arise. Some examples in this regard are areas with smoke from fireplaces, campfires, wildfires, near industrial areas and busy roads from urban zones, and during calm weather, when pollutants are accumulating in specific areas, according to the local geography (for example, hot days). Heavily trafficked streets set between continuous rows of high buildings, named canyon streets, lead to urban pollution hotspots caused by very high levels of traffic generated air pollutants and very restricted atmospheric dispersion.

Particulate matter has a small size; consequently, this can enter inside buildings, thus generating a high indoor particle pollution level. Fine particulate matter pollution can have a seasonal impact. For example, in cool weather, fine particle nitrates are more likely to be formed, as well from wood stove and fireplace use. Therefore, in the mountains, especially in the wintertime, when the wood is burned to release heat, pollution particle level is high. Another consideration for the polluted areas is represented by the geographical aspects of the area, such as valleys, hills and more [7]. A broad understanding of the movement and dispersion of the pollutants in the atmosphere is described in Fig. (**1.3**).

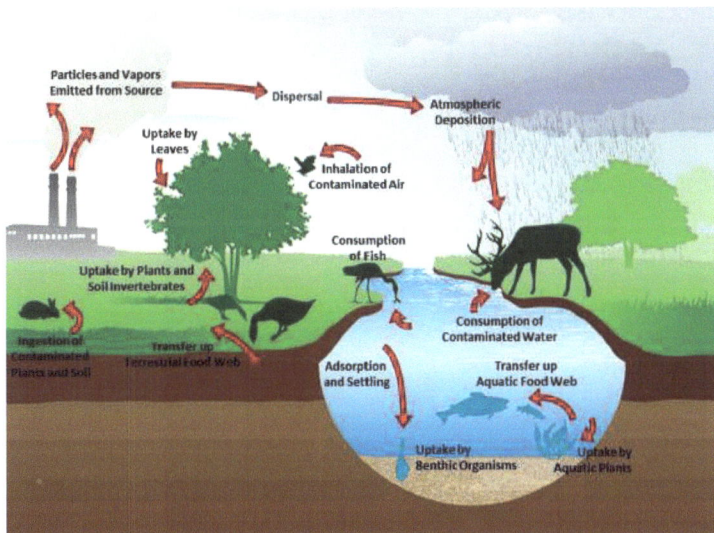

Fig. (1.3). Scenario for air particles emission and dispersion in the environment (epa.gov).

A mixture of solids and liquids, including carbon, organic chemicals, sulfates, nitrates, mineral dust, and water suspended in the air, makes up Particulate Matter (PM). PM has different sizes. The most dangerous are PM_{10} and $PM_{2.5}$; 10 and 2.5 refer to the diameter of the particles, having a micrometer scale (μm). This type of pollutant is emitted by industries, building works, burning diesel and petrol engines, dust from roads, *etc.* Diesel vehicles produce more PM than petrol vehicles [8].

Nitrogen dioxide (NO_2) is a gas resulting from burning fuel from vehicles, power plants or heating units. In the case of big cities, the major source of NO_2 releases into the atmosphere is diesel vehicles.

Three oxygen atoms combine to form the gas ozone (O_3). Ozone is found in the upper atmosphere of the Earth and is responsible for filtering out harmful ultraviolet radiation emitted by the sun. Ozone at the ground level is obtained through a chemical process between sun rays that come in the reaction with organic gases and nitrogen oxides released from the cars' burning fuels, power plants, *etc* [8].

Carbon Monoxide (CO) is an odorless gas and is resulted after burning. The greatest contributors to the CO amount are vehicles which burn fossil fuels. Carbon monoxide is dangerous to humans' health at the level of concentration [8].

In addition to the negative effects on human health, the pollutants described above, once released into the atmosphere, can result in changes in climate change. According to the Eurostat, in 2019, Greenhouse Gas Emissions (GHG) in the EU were about 1 billion tonnes of CO_2 (lower than in 1990). The largest reductions in GHG emissions were observed in Latvia, Estonia, Romania and Lithuania [9]. Fig. (**1.4**) represents the GHG emissions by source in Romania, Spain, and Italy:

Soil Pollution

Soil represents a complex resource used for many purposes:

- Food and biomass production;
- Storage, filtering and transformation of substances, such as water, carbon, nitrogen, *etc.*;
- Supply of raw materials;
- Regulating air quality.

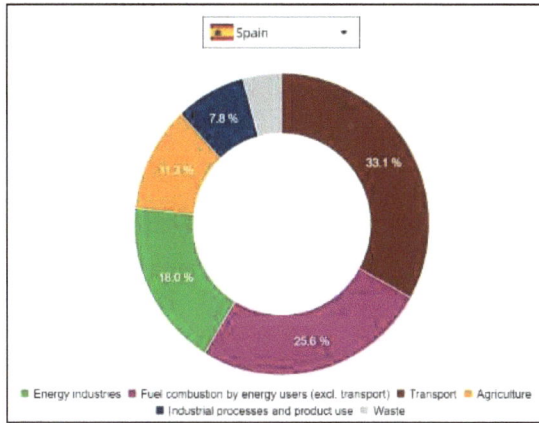

(a)

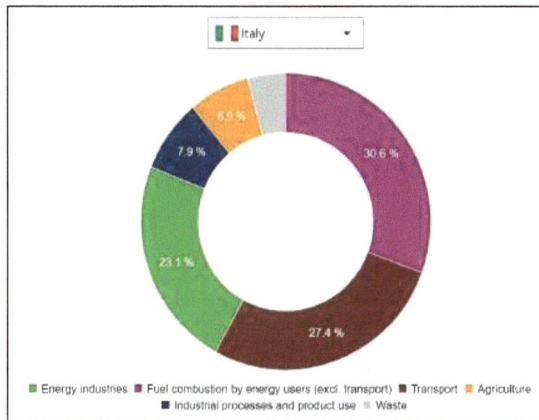

(b)

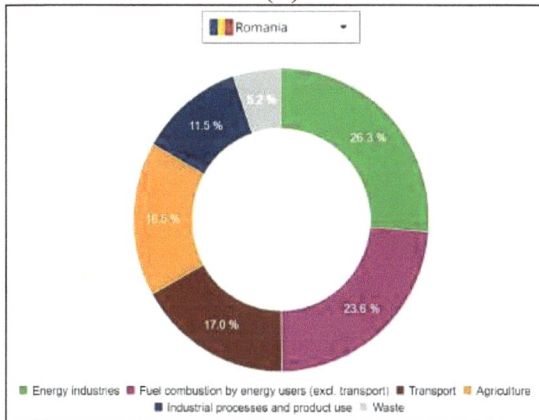

(c)

Fig. (1.4). Greenhouse gas emissions considering the different sources in 3 countries, according to EURO-STAT (Source: https://ec.europa.eu/eurostat/cache/infographs/energy/bloc-4a.h tml?lang=en).

Soil pollution can severely degrade ecosystem services and can affect human health. Also, processes like leaching and run-off can move the pollutants from soil to groundwater and surface water, which are vulnerable to pollution. In the same way, the dynamics of soil pollutants mobilization can affect air quality. Persistent organic pollutants accumulated in soils can be remobilized by the volatilization process and affect the atmosphere at local and regional levels.

The transport of pollutants *via* air-soil-water systems (Fig. **1.5**) is difficult to analyze. The understanding of the severity and extent of the pollution impact needs modern technologies and expensive equipment for measuring and monitoring the extent of pollution.

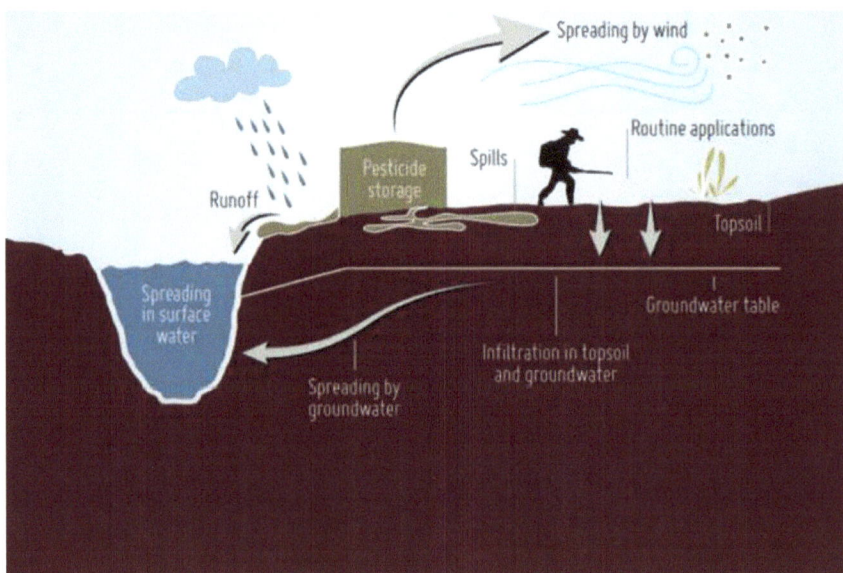

Fig. (1.5). Transport pathway of soil pollutants in the environment (Source: FAO, 2000).

Soil pollution occurs when persistent toxic substances are added to the medium and change the physical, chemical, and biological properties of the soil and reduce its fertility. Soil pollution occurs due to the illegal dumping of industrial wastes (which contains lead, cadmium, copper, acids, cyanides, *etc.*) or radioactive waste, overusing of pesticides (insecticides, algicides, *etc.*) and fertilizers, deforestation, and soil erosion. Soil erosion as well may be part of a natural process, however, in most cases, it is accelerated by human activities. Organic chemicals such as fertilizers, pesticides and other industrial/agricultural chemicals, once released in nature, are sorbed by the particulate surface, thus associating the chemicals with the soil structure and particles. This process contributes to lowering the soil quality and accelerates erosion [1].

Another source of man-made chemicals which contributes considerably to soil pollution is the emission of pollutants eliminated in the atmosphere (SO_2, NO_x, CO *etc.*) and delivered to the soil surface through wet and dry depositions.

Soil properties can be affected by point-source pollution (former factory sites, inadequate waste and wastewater disposal, uncontrolled landfills, excessive application of agrochemicals, leakage from tank installations, *etc.*) and diffuse pollution (nuclear power and weapons activities, uncontrolled waste disposal and contaminated effluents released in and near catchments, land application of sewage sludge, the agricultural use of pesticides and fertilizers, excess nutrients and agrochemicals that are transported downstream by surface runoff, flood events, *etc.*).

Main pollutants in the soil, usually originating from anthropogenic activities, are represented by: heavy metals, nitrogen and phosphorus, pesticides, and persistent organic pollutants.

Heavy metals (As, Pb, Cd, Cu, Hg, Sb, Se, Sn, Zn) naturally occur at low concentrations in soils, but at high concentrations, may cause phytotoxicity and harm human health. They are persistent and complex pollutants because of their non-biodegradable nature, which causes them to readily accumulate in tissues and living organisms. The main anthropogenic sources of heavy metals are industrial areas, mine tailings, disposal of high metal wastes, leaded gasoline and paints, application of fertilizers, animal manures, sewage sludge, pesticides, wastewater irrigation, coal combustion residues, spillage of petrochemicals, and atmospheric deposition from varied sources [20].

Nitrogen (N) and phosphorus (P) are essential components of all living organisms. Soil productivity is the result of several factors, such as: soil fertility, good soil management practices, availability of water supply, and suitable climate. A short definition of soil fertility refers to soil rich in nutrients (nitrogen, phosphorus, potassium) in an available form for plants. Agricultural soils can be rich in nutrients or deficient in nutrients. In this last case, an optimum amount of essential nutrients N, P and K through synthetic fertilizers is needed to be applied to sustain crop production.

Nitrogen and phosphorus become pollutants when they are applied in excess to agricultural soils in the form of fertilizers or in areas of intensive livestock production. The intensification of agriculture for global food security leads to the excessive application of fertilizers designed to rapidly and cheaply grow plentiful crops and raising large numbers of farm animals.

These nutrients are able to leach into the groundwater or be transported to surface water bodies by runoff, causing eutrophication or leading to high nitrate concentrations. Pollution problems appear when fertilizers are applied in excess without considering the main factors that influence the nutrients leaching: soil texture, meteorological conditions, irrigation schedule, the vegetation period of the plants *etc.*

In humid northern temperate zones, which are primarily covered by coniferous forests, and humid tropics, which are covered by savannah and tropical rain forest, acid soil is defined as soil with a low pH value (less than 5.5). Natural acid soil occurs due to excessive weathering of soil minerals, resulting from high rainfalls, and raised temperatures. However, acid soil may be developed because of contamination with different acid fertilizers used in agriculture. The most common acid forming fertilizers used in agriculture are ammonia and urea.

On agricultural lands, pesticides are applied to reduce crop losses due to insect pests, weeds and pathogens. In the present context (climate change, extreme events, pollution impact on biodiversity, natural resource depletion, global food security), the pesticide application is necessary in terms of increased production of food and fibre, and amelioration of vector-borne diseases. Pesticides can contaminate soil, water, air, and other vegetation (Fig. **1.6**) when they are applied in higher amounts than needed, using practices that contribute to their spreading into the environment, such as spraying with not suitable/not maintained/not calibrated application equipment or by planes into vast regions, affecting inhabitants and non-target organisms.

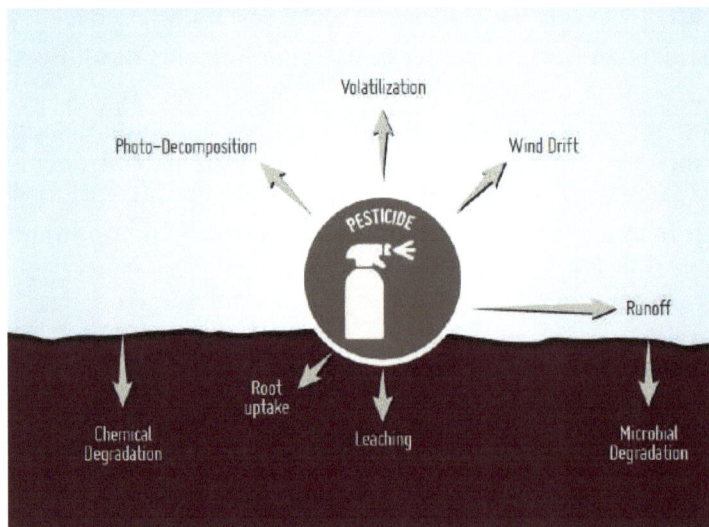

Fig. (1.6). Behaviour of pesticides in the environment (Source: Singh, 2012).

Pesticides can be either organic or inorganic synthetic molecules, and their persistence, behaviour and mobility in the environment are extremely varied.

Persistent organic pollutants (POPs) are chemical substances that persist in the environment, enter the food chain by accumulating in the body fat, becoming more concentrated as they move from one organism to the next, and also have high mobility in the environment.

They have been used in agriculture, disease control, manufacturing and many industrial processes, but their use and production have significantly been reduced since the adoption of the Stockholm Convention in 2018.

POPs are highly resistant to degradation and persist for a long time in the environment. Soils are the main environmental sink for these pollutants, but they can easily penetrate water in its gaseous phase during warm weather and volatilize from soils into the atmosphere.

Persistent organic pollutants present great affinity to organic matter and lipid membranes of cells, so the pollution by POPs has adverse effects on human health and the environment.

Water Pollution

Water pollution occurs when toxic compounds enter the water, that degrade its quality and make the water not suitable for drinking or other purposes, like irrigation, domestic and industrial use, and recreation. Groundwater/surface water contamination, most common, is with hazardous organic chemicals and has a high risk to human health and the environment. Major classes of hazardous organic chemicals which are immiscible, and are considered with high significance in water pollution, are chlorinated solvents, coal tar/creosote, hydrocarbon fuels, and polychlorinated biphenyls. Immiscible contaminants are very hard to be removed from the water and require a great cost and time for site remediation. Water contamination usually occurs in conditions of transport, storage, use and disposal of hazardous chemicals [1]. EPA has elaborated a good example of the connection between household drinking water and nearby working industry that releases intentionally and unintentionally polluted water, as illustrated in Fig. (**1.7**).

Surface water and groundwater systems are connected in most landscapes. Groundwater and surface water physically overlap at the groundwater/surface water interface through the exchange of water and chemicals, so the pollution of surface water influences the groundwater quality and is reciprocal.

Sources of Ground Water Contamination

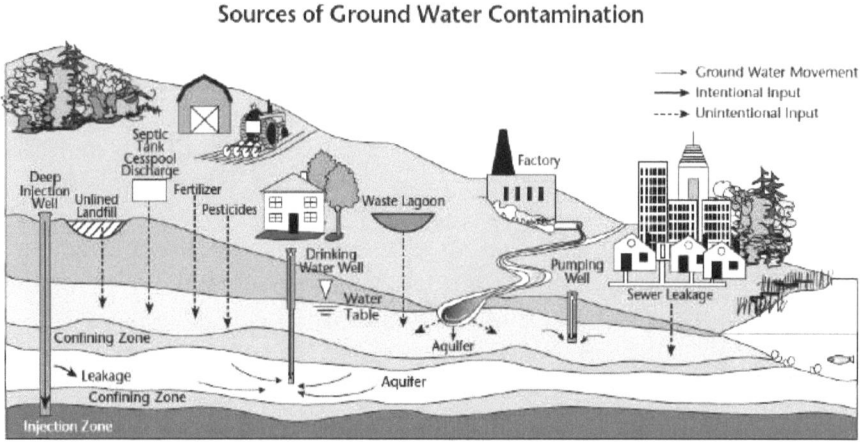

Fig. (1.7). Example of sources for groundwater contamination (EPA, 1998) [1].

Inadequate management of urban, industrial and agricultural wastewater means the drinking water of hundreds of millions of people is dangerously contaminated or chemically polluted. According to European legislation, the wastewater must be treated prior to disposal in the environment in order to protect both public health and groundwater and surface water quality. There are three levels of wastewater treatment:

- Primary treatment: it includes physical processes and removes about 60% of total suspended solids; dissolved impurities are not removed;
- Secondary treatment: it removes the soluble organic matter that escapes primary treatment (more than 85% of total suspended solids) and it is usually accomplished by biological processes;
- Tertiary treatment: remove more than 99% of all the impurities from sewage, producing an effluent of almost drinking-water quality.

The proper configuration of the wastewater treatment plant is selected according to the governmental standards, depending on local environmental conditions, source of wastewater and subsequent use of the effluent.

High nutrient concentrations (nitrate and phosphate) in surface water conduct to eutrophication of water bodies. This phenomenon causes an intense growth of algae that reduces oxygen levels in the water and many organisms, such as fish, amphibians and water insects, can no longer survive. Also, the eutrophication has negative consequences for drinking water sources, fisheries, and recreational activities.

A direct consequence of fossil fuels burning industries, which release carbon dioxide, nitrogen oxides and sulfur dioxide gases into the atmosphere, is represented by the acidification of water bodies. These pollutants land in water bodies directly or, more often, mix with water in the atmosphere before falling as acid rain and impacting the water bodies' chemistry. As a consequence, marine life is affected by the acidification of aquatic systems.

Playing a key role in the Earth's climate and weather systems, unique habitats, and material and recreational ecosystem services that contribute to human well-being, ocean acidification represents a major issue. The most effective way to limit ocean acidification is to implement solutions to reduce the use of fossil fuels.

Improper waste disposal can damage the quality of surface and groundwater bodies by changing their chemical, physical and organoleptic properties. This process raises the toxicity of the water with consequences on aquatic biodiversity, making freshwater unsafe for human consumption and any body of water improper for recreational activities and tourism. Citizens' education in waste management is very important. Knowledge about waste prevention and sustainable disposal, and its impact on health and the environment has vital importance to safe water.

ENVIRONMENTAL POLLUTION AND HEALTH EFFECTS

Over the past centuries, due to rapid industrialization, scientists have focused on air, water and soil pollution and their negative impact on human health. Climate change, environmental pollution, and depletion of the ozone layer lead to multiple risks to human health and the ecosystem.

A study involving 27 EU members made by the Environmental Protection Agency states that "*in 2019, approximate 307,000 premature deaths were attributed to chronic exposure to fine particulate matter present in the air, 40,400 premature deaths were attributed to chronic nitrogen dioxide exposure; 16,800 premature deaths were attributed to acute ozone exposure*" [10]. However, the statistics shows that in 2019 compared to 2018, the number of premature death cases has decreased, most probably as a consequence of applying the regulations imposed by the national regulations in force in accordance with the Green Deal and by Zero Pollution Action Plan [11].

A number of documents and regulations have been proposed as part of the countries' ambitious efforts to build a Healthy Planet in order to track, report, prevent, and address air, soil, and water pollution issues. However, despite the effort and strategies to control the pollution, no immediate actions are applied at the local level by the managers of the industries, farms, *etc* [11].

Health effects due to air pollution are worldwide recognized. It represents the major risk factor for human health, leading to respiratory and cardiovascular diseases. The most persistent and dangerous of air contaminants is particulate matter, that is responsible for heart diseases (as represented in Fig. (**1.8**)), chronic pulmonary diseases, stroke, lung cancer and many other respiratory diseases [12].

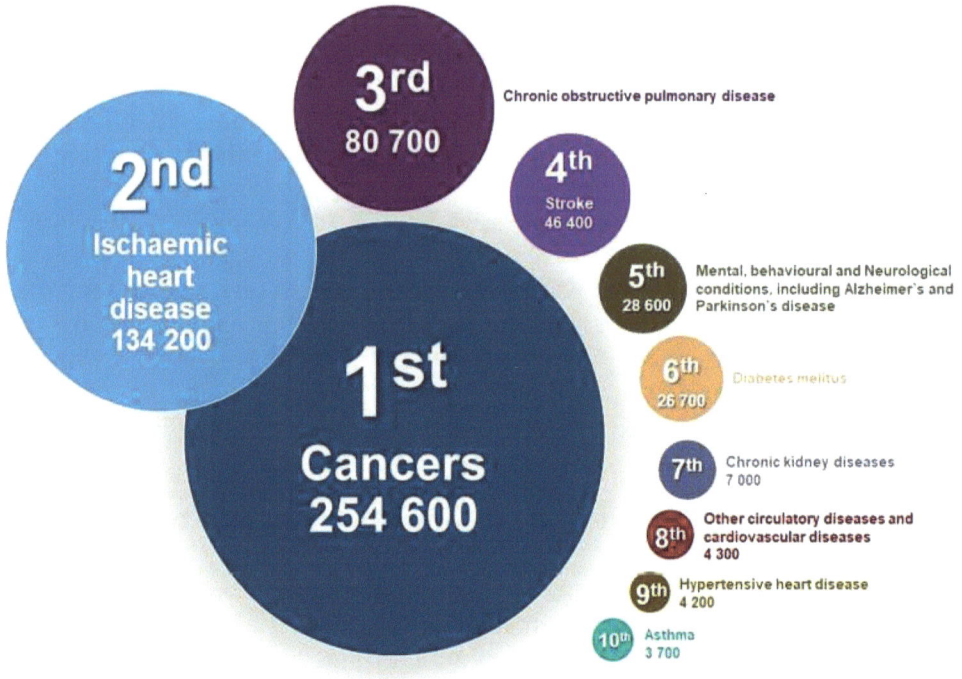

Fig. (1.8). First 10 most likely cases of the diseases resulted from environmental pollution [13].

Pollutants in the environment contribute to a lower quality of life because they cause people to develop health problems such as heart disease, diabetes, and cancer as a result of constant exposure. Air pollution, based on the evidence provided by the EEA, has the most significant impact on health, leading to 400 000 premature deaths per year. Noise pollution is the second type of pollution that drives over 12000 premature deaths per year worldwide [12].

Fig. (**1.9**) presents the most common air pollutants and their effects on human health developed by the European Environment Agency. It is worth noting that exposure to particulate matter causes the most harm to human health.

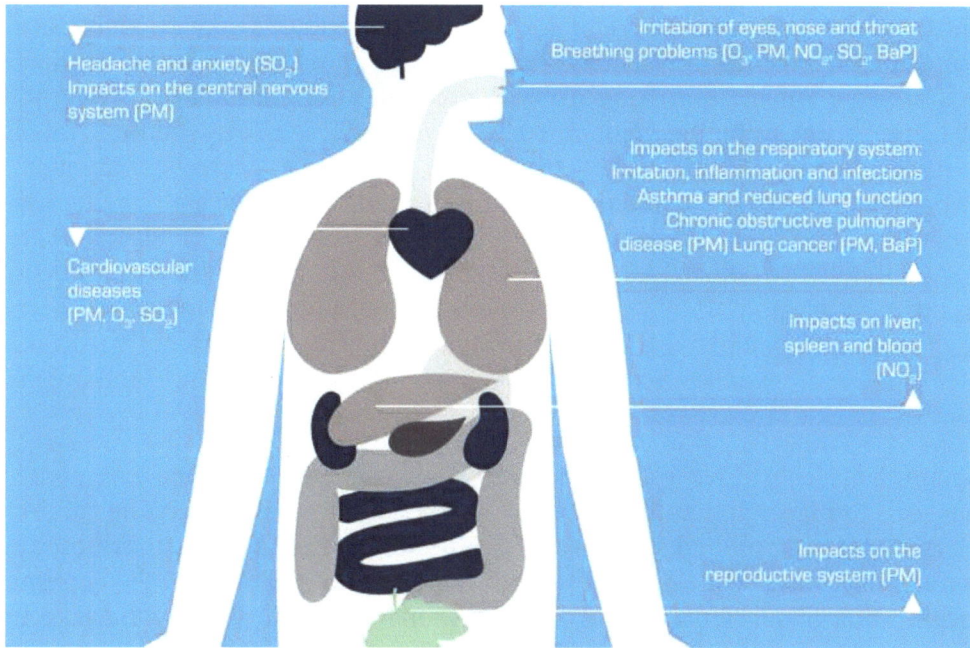

Fig. (1.9). Most common air pollutants and their effects on the human body (Source: European Environment Agency [14]).

As previously stated, exposure to air pollutants can occur both indoor and outdoor. Scientists discovered that increased exposure to air pollutants increases the risk of severe COVID-19 complications due to the impact of pollutants in causing respiratory and cardiovascular diseases. As a result, people who are constantly exposed to PM are more vulnerable to the effects of COVID-19 [12, 15]. Herein it was developed a model for a better understanding of the link between humans and environmental pollution. According to the World Health Organization, in 2020, 5.8 billion people used safely managed drinking-water services (improved water sources located on premises, available when needed, and free from contamination). The remaining 2 billion people without safely managed services included:

- 1.2 billion people with basic services, meaning an improved water source located within a round trip of 30 minutes,
- 282 million people with limited services, or an improved water source requiring more than 30 minutes to collect water,
- 368 million people taking water from unprotected wells and springs, and,
- 122 million people collecting untreated surface water from lakes, ponds, rivers and streams.

Contaminated water and poor sanitation expose individuals to preventable health risks: cholera, diarrhoea, dysentery, hepatitis A, typhoid, and polio. Also, the use of nitrate-contaminated drinking water to prepare infant formula is a well-known risk factor for infant methemoglobinemia, known as blue baby syndrome, this is a condition where a baby's skin turns blue due to a decreased amount of hemoglobin in the baby's blood. At the european level, the Nitrate Directive aims to reduce water pollution caused or induced by nitrate from agricultural sources and requires EU Member States to monitor the quality of waters and to identify areas that drain into polluted waters or are at risk of pollution. The World Health Organisation recommends a maximum admissible level of nitrate in drinking water of 50 mg/l, which is needed to protect infants, young children, and pregnant women.

Contaminants can move relatively freely between environmental components, from soils to plants and animals, to the atmosphere, to the water bodies and *vice versa*. Therefore, polluted soil poses a risk to human health. Safe, nutritious, and good-quality food can only be produced if the soils have the optimum concentration of nutrients. A good fertilizer management plan should take into consideration a lot of factors that increase nutrient leaching potential (the amount of applied fertilizer, soil texture, precipitation and irrigation schedule, crop management, *etc.*).

Exposure-disease model is a short concept/diagram showing the path of the pollutant to the human, as illustrated in Fig. (**1.10**).

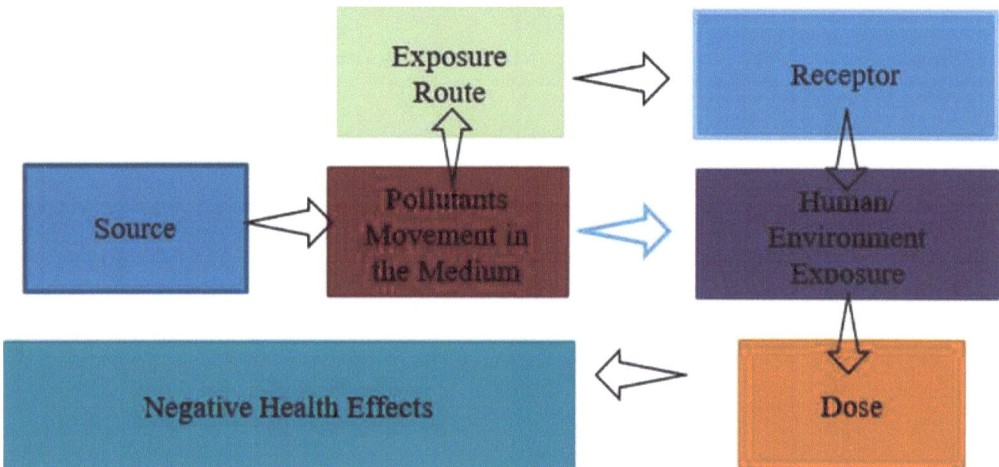

Fig. (1.10). Exposure-Disease Model (Source [16]).

Source

The exposure-disease model proposes a path from exposure to the subsequent development of a negative health outcome. The exposure-disease model would begin with sources and emissions. The sources of exposure are the locations/events where the contaminant was first released into the environment and caused harm. Such sources may include:

- Natural sources
- Agricultural practices
- Consumer products
- Industrial emissions
- Vehicle emissions
- Waste collection and management [17].

Environmental Media

The media through which humans/animals are exposed to a contaminant/agent is referred to as environmental media. There are primarily three media (air, water, and soil) through which the contaminant can move from the original source, depending on the properties of the pollutant and the source [18]:

- Biota (plants and animals)
- Air
- Soil/Sediment
- Water *(e.g.,* groundwater, surface water)
- Additional environmental compartments that may contain contaminants

Exposure Routes

Environmental media is the medium through which humans/animals are exposed to a contaminant/agent. There are primarily three media (air, water, and soil) through which the contaminant can move from its original source, depending on the properties of the pollutant and the source.

This exposure-disease model concept now assists scientists in calculating and determining the likelihood of human risk following the intake of certain contaminants and their effects on human health. This strategy is used all over the world to demonstrate the path of a pollutant to the human body, as shown in Fig. **(1.11)** [19, 20].

Fig. (1.11). Routes of human exposure to the contaminants (Source [17]).

CONCLUDING REMARKS

Environmental pollution can occur from natural or anthropic sources. The main natural pollution sources are wildfires and volcanic eruptions, while the anthropic activities that have the most important impact regarding environmental pollution are represented by burning of the fuels, untreated industrial and municipal wastewater discharge, uncontrolled dumping of waste, excessive applications of fertilizers and pesticides, accidental spills of toxic substances. Pollutants are able to move from one environment system (soil, water, air) towards another through migration processes, so the air pollution can affect also the water and soil quality, and reciprocal. The influence of environmental pollution on human health and biodiversity is established by Environmental or Ecological Risk Assessment Studies. The literature in this field highlights that the environmental pollution represents a major risk factor for human health leading to health problems such as respiratory and cardiovascular diseases, diabetes, cancer, *etc.* Exposure to different pollutants affects especially the population with pre-existing diseases, older people, and children. Other negative consequences of environmental pollution include material, financial, cultural and recreational losses generated by floods, drought, soil erosion, sea level rise, *etc.* Also, pollution is one of the key drivers of habitat fragmentation and destruction, behavioural changes in species, global biodiversity loss, *etc.* Human activities that conduct to environmental pollution are necessary, so we cannot stop them, but we can use best available techniques developed to prevent and control industrial pollution, and thus ensure a high level of human health and environmental protection. At the same time, the population education regarding environmental protection is very necessary to sustain a clean nature.

REFERENCES

[1]   I. L. Pepper, C. P. Gerba, and M. L. Brusseau, Environmental and Pollution Science vol. 53. 3rd Ed. Elsvier, pp. 617-633, 2019.
[http://dx.doi.org/10.1016/C2017-0-00480-9]

[2]   Y. Haseli, "Fundamental concepts", *Entropy Anal. Therm. Eng. Syst,* pp. 1-11, 2020.
[http://dx.doi.org/10.1016/B978-0-12-819168-2.00001-5]

[3]   T.P. Curtis, "Low-energy wastewater treatment: strategies and technologies", *Environmental*

microbiology, 8;2, 2010.

[4]   S. Telles, S. K. Reddy, and H. R. Nagendra, Environment vol. 53. SHANKAR IAS Academy, no. 9, p. 436, 2019.

[5]   WHO, "Chronic obstructive pulmonary disease (COPD)", Available at: https://www.who.int/news-room/fact-sheets/detail/chronic-obstructive-pulmonary-disease-(copd) (Accessed on: 2022).

[6]   "European environmental protection agency, european city air quality viewer.european city air quality viewer", Available at: https://www.eea.europa.eu/themes/air/urban-air-quality/european-city-air-quality-viewer (Accessed on: 2022).

[7]   "U. epa (environmental protection agency, what is particle pollution? Particle Pollution and Your Patients Health", Available at: https://www.epa.gov/pmcourse/what-particle-pollution (Accessed on: 2022).

[8]   "Earth system research laboratories global monitoring laboratory, the noaa annual greenhouse gas index (AGGI).the NOAA annual greenhouse gas indEX (AGGI)", Available at: https://gml.noaa.gov/aggi/aggi.html (Accessed on: 2022).

[9]   "EUROSTAT how are emissions of greenhouse gases in the EU evolving?", Available at: https://ec.europa.eu/eurostat/cache/infographs/energy/bloc-4a.html?lang=en (Accessed on: 2022).

[10]   "European environmental agency health impacts of air pollution in europe.air quality in europe 2021 health impacts of air pollution in europe", Available at: https://www.eea.europa.eu/publications/air-quality-in-europe-2021/health-impacts-of-air-pollution (Accessed on: 2022).

[11]   UNEP, "United Nations Environment Assembly of the United Nations Environment Programme Strategy for Private Sector Engagement," pp. 7–11, 2019.

[12]   "European environment agency", *EEA Report: Healthy environment, healthy lives: How the environment influences health and well-being in Europe.*, no. 21, 2020.

[13]   "European commision, communication from the commission to the european parliament, the council, the european economic and social committee and the committee of the regions empty. pathway to a healthy planet for all", Available at: https://eur-lex.europa.eu/legal-content/EN/TXT/?uri=CELEX%3A52021DC0400 (Accessed on: 2022).

[14]   "OECD/European Union: Air pollution and its impact on health in Europe: Why it matters and how the health sector can reduce its burden", In: *Heal. a Glance Eur. 2020 State Heal.* EU Cycle, 2020, pp. 83-109.

[15]   D.M. Cocârţă, M. Prodana, I. Demetrescu, P.E.M. Lungu, and A.C. Didilescu, "Indoor air pollution with fine particles and implications for workers' health in dental offices: A brief review", *Sustainability*, vol. 13, no. 2, p. 599, 2021.
[http://dx.doi.org/10.3390/su13020599]

[16]   K. Jones, "Human biomonitoring in occupational health for exposure assessment", *Port. J. Public Health*, vol. 38, no. 1, pp. 2-5, 2020.
[http://dx.doi.org/10.1159/000509480]

[17]   "University of boston, exposure assessment introduction to basic concepts.exposure assessment introduction to basic concepts", Available at: https://sphweb.bumc.bu.edu/otlt/mph-modules/expo-sureassessment/exposureassessment3.html(Accessed on: 2022]

[18]   "EPA exposure assessment tools by routes : Ingestion conducting an ecological risk assessment", Available at: https://www.epa.gov/risk/conducting-ecological-risk-assessment(Accessed on: 2022).

[19]   "EPA environmental protection agency, exposure assessment tools by routes : Ingestion", Available at: https://www.epa.gov/expobox/exposure-assessment-tools-routes-ingestion(Accessed on: 2022).

[20]   N. Rodríguez-Eugenio, M. McLaughlin, and D. Pennock, *Soil Pollution: A hidden reality.* FAO: Rome, 2018, p. 142. Available at: https://www.fao.org/3/I9183EN/ i9183en.pdf.

CHAPTER 2

Basic Concept of Environmental Risk Assessment

Diana Mariana Cocârță[1,*] and **R. Lupu**[1]

[1] *University POLITEHNICA of Bucharest, Faculty of Energy Engineering, Splaiul Independentei 313, RO-060042 Bucharest, Romania*

Abstract: Rapid growth and expansion of the chemical and energy industries have led to an increase in chemical emissions and the potential for accidents, such as fires, explosions, and spills. These potential consequences have caused concern among industries and regulators, leading to an interest in understanding the risks associated with these emissions into the environment and accidents. This knowledge is crucial for complying with laws and regulations as well as for reducing adverse effects on human health and the environment. Consequently, this chapter intends to introduce readers to the Environmental Risk Assessment and the steps that should be accounted for the successful results, representing an important study on understanding the risks and minimizing the negative effects as much as possible. The first part of the chapter is dedicated to presenting specific knowledge on Risk and Hazard, evidencing differences between the two terms. The second section emphasizes the value of human health risk assessment in determining the potential effects of a hazard on the health of an individual, a group of individuals, or a community, while simultaneously outlining the steps that must be taken for the Ecological Risk Assessment.

Keywords: Ecological Risk Assessment, Hazard, Human Health Risk Assessment, Risk Assessment.

INTRODUCTION

The concept of risk assessment first appeared in the 1970s, with its origins in the United States, because of addressing issues related to human health, particularly the approach needed to be more balanced and efficient. In response to the proposal, a more consistent risk assessment program was implemented in the United States in the 1970s and 1980s. As a result, various governments adopted the risk assessment procedure, while also developing different subdivisions of risk assessment, such as environmental risk assessment and human risk assessment [1].

---

[*] **Corresponding author Diana Mariana Cocârță:** University POLITEHNICA of Bucharest, Faculty of Energy Engineering, Splaiul Independentei 13, RO-060042 Bucharest, Romania; E-mail: dianacocarta13@yahoo.com

Due to the high interest in contaminants in relation to ecological processes and effects, the Environmental Protection Agency of the United States began to develop guidelines on ecological risk assessment in the 1980s. In 1983, the US National Research Council formalized the risk assessment model in the "Red Book," and as a result, the terms and procedures of risk assessment and risk management were clearly differentiated [1]. Later, similar documents for assessing ecological risk were developed in Europe, Canada, and Australia.

UNDERSTANDING OF RISK

The four steps of risk assessment are hazard identification, dose-response assessment, exposure assessment, and risk characterization, which serve as the foundation for today's risk assessment research.

Because of the high impact of stressors, risk assessment has recently evolved significantly. Risk assessment methods are now used in a variety of areas, including ecological risk assessment, health risk assessment, life-cycle risk assessment, and others.

Ecological risk assessment, which refers to environmental issues, includes significant information on cumulative risk assessment that was not previously considered. The integration of multiple stressors, management options, and endpoints in the developed conceptual model is referred to as cumulative risk assessment. Aside from that, various countries have developed their own environmental risk assessment software, which has successfully integrated human health risk assessment into the system. Programs can integrate and output information based on the information provided in the data input. Ecological risk assessment has recently been used in the study of genetically modified organisms, generating useful results [2].

i. What is Risk?

According to the Oxford Dictionary, the risk is defined as "an event/situation that involves exposure to danger," and it is also used to describe what is likely to happen (in the future) if the required preventive measures are not carried out in accordance with the standards. In other words, risk was defined by the Royal Society in 1992 as "*the combination of the probability or frequency of occurrence of a defined hazard and the magnitude of the consequence of the occurrence*" [3]. However, there are situations when the danger cannot be predicted or may have a small percentage of circumstance. Herein, the risk assessment study is performed to identify the risk and propose solutions for its elimination.

The risk assessment mechanism is applied in a wide range of domains and academic disciplines. Engineers, for example, use the risk assessment study process to determine the likelihood and effects of component failure while designing a bridge [4]. Engineers were primarily involved in the construction of churches during the Romanesque to Gothic transition stage. Many structures collapsed, and deaths and environmental damage occurred during that time. These events aided society in evolving over the centuries and implementing efficient techniques to reduce the risks of collapses and loss of life. To avoid negative events, engineers and organizations now conduct rigorous risk assessments. Risk assessment has become a popular method for investigating environmental issues and assessing various types of risks [3].

According to the EPA standards, the risk is understood based on the answers to the following questions:

1. **How much of a pollutant exists** in the environmental matrix *(e.g.,* soil, water, air) and who is affected by it?

The answer to this question focuses on identifying potential risks and goods that may be impacted by them (humans, animals, environment, *etc*).

2. **How much contact (exposure)** does an individual or an ecological receptor have with the contaminated medium?

In this case, it is thought to continue with the analysis to determine the causes and potential amount of the asset that was exposed to the stressor/contaminant.

3. **How it affects and what are the consequences** on human health and the ecosystem [5].

It is determined here how the stressor affects the medium of exposure (people, environment), as well as whether the risk has the potential to cause harm or an adverse consequence. In other words, risk is expressed as the result of two factors: the likelihood of exposure and the adverse consequence.

$$Risk = probability\ of\ exposure \times \text{adverse consequence} \qquad \textbf{(2.1)}$$

For example, smokers are more likely to develop lung cancer:

"People that smoke cigarettes are 20 times (example) more likely to develop lung cancer than non-smokers".

Risk is expressed as a probability or likelihood of an injury to happen, while the hazard refers to the responsible agent for this cause *(e.g.,* smoking cigarettes) [2].

UNDERSTANDING OF HAZARD

Hazard, on the other hand, is defined as "the potential to cause harm". A hazard is defined as a substance that is chemical, physical, or biological in nature and has the potential to harm human health if the contaminant is present in the environment. The intensity of the consequences that may result defines the nature of the substance (hazard) present in the environment.

The distinction between hazard and risk can be illustrated by the following example: in a chemical laboratory, there may be a large quantity of hazardous chemicals, some of which are acids. Acids can harm humans only when they are exposed to them, and the extent of harm depends on the properties of the substance. In other words, if a human comes into contact with diluted acid, the risk is reduced, but the hazard remains [1].

If a hazard occurs, a simple way to describe it - is to use the probability scheme. Scientists can estimate the likelihood of the hazard and how many people are affected using the following method, which considers a range of probability and attributed description (Table **2.1**).

Table 2.1. The likelihood of the hazard to occur in a certain situation (Source: Telus, Draft from Environmental Impact Statement) [6].

Likelihood	Description	Probability	Mid Interval	Community Outlook
Eliminated	Would not occur as a result of being designed out of the program	P 0	0.00	Not affected
Remote	May occur only in exceptional circumstances	$0.01<P<0.10$	0.05	Few or no people affected
Unlikely	Could occur at some time	$0.11<P<0.40$	0.25	Some people affected
Possible	Might occur at some time	$0.41<P<0.60$	0.50	Many people affected
Likely	Will probably occur in most circumstances	$0.61<P<0.90$	0.75	Most people affected
Almost Certain	Is expected to occur in most circumstances	$0.91<P<1.00$	0.95	Almost everyone affected

Following the estimation method, the obtained results can be integrated and used to evaluate and comprehend the hazard's impact. Consequently, information about the geographic scope, timeframe, ecological and social sensitivity, and potential cumulative effects is required to determine the consequence of the hazard. Table **2.2** displays possible consequence categories, which allow us to visualize the impact from various perspectives [6].

Table 2.2. Visualizing the consequence of the hazard (Source: Telus, Draft from Environmental Impact Statement) [6].

Consequence Descriptor Probability					
-	**Health**	**Environmental**	**Financial Loss**	**Project Delivery**	**Social**
Insignificant	No Injuries	None	Low financial loss	Trivial	Insignificant
Minor	First aid treatment	On-site release immediately contained	Medium financial loss	Project can be completed with changes	Additional local engagement
Moderate	Medical treatment required	On-site release contained with outside assistance	High financial loss	Project can be completed with moderate changes	Additional meetings
Major	Extensive Injuries	Off-site release with no detrimental effects	Loss of production capability major financial loss	Project can be completed with moderate changes (redesign)	Reactive media plan, recovery plan, working committees
Catastrophic	Death	Toxic release off-site with detrimental effect	Cessation of production capability/ Huge financial loss	Project incapable of completion/ Unviable	No social licence to operate

HAZARD AND RISK

Hazard and risk are frequently confused as synonyms, but their meanings are different. As previously stated, a hazard is something that has the potential to cause harm, and the risk is the likelihood of the hazard causing harm. These two terms are the most important factors used in risk assessment methodologies and subsequent studies [7], which will be widely described in the following chapters.

For a better understanding of the difference between hazard and risk, the European Food Prevention Council has created an intuitive illustration (Fig. **2.1**).

As illustrated in Fig. (**2.2**), these two terms are widely used in the three activities of risk analysis, risk evaluation, and risk control. The basic studies performed in the Risk Assessment are Risk Analysis and Risk Evaluation. All three are at the heart of the risk management decision-making process.

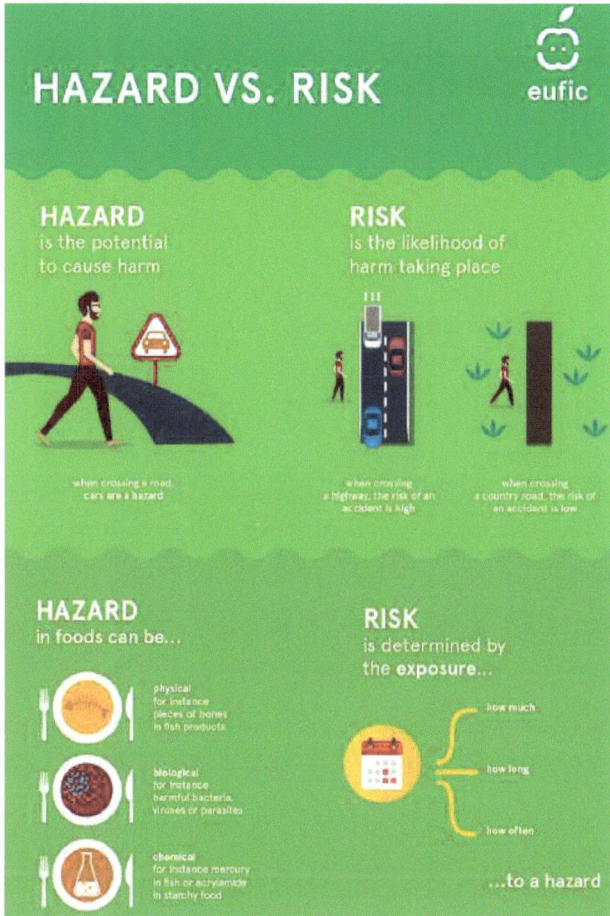

Fig. (2.1). Understanding the difference between hazard and risk (Source: https://www.eufic.org/en/understanding-science/article/hazard-vs.-risk-infographic).

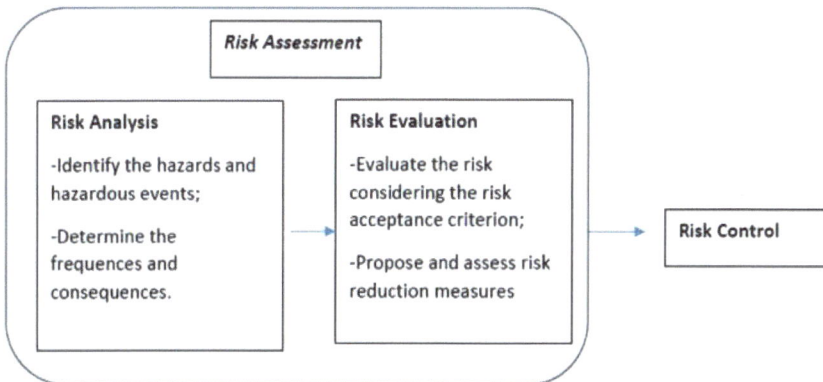

Fig. (2.2). Risk assessment phases [7].

Given the preceding phases, it is critical to understand what a Risk Assessment is and how Risk Analysis and Risk Evaluation aid in developing a strategy to manage any potential problems for the environment and human health.

Risk Assessment is defined as the process of determining the consequences and problems that may occur of/ produce a risk. Risk assessment is a tool for identifying existing hazardous situations, anticipating potential problems, establishing priorities, and providing a foundation for regulatory controls and corrective actions. Risk assessment can be defined as the study of estimating the likelihood and magnitude of an event while taking economic, health/safety, and environmental factors into account.

Risk Analysis is a procedure that examines the consequences of a risk that has occurred. A risk assessment study is usually followed by a risk management study, which reviews the information gathered from risk assessment and risk analysis to reduce or eliminate the impact of the occurred risk [8].

Risk Evaluation is the procedure by which a risk is evaluated based on risk acceptance criteria, and risk reduction measures are generally proposed at this stage.

There are many different types of Risk Assessment studies, but in this chapter, we will concentrate on Human and Ecological Risk Assessment [9].

HUMAN HEALTH AND ENVIRONMENTAL RISK ASSESSMENT

Environmental Risk Assessment or *Ecological Risk Assessment* (ERA) examines the probability that ecological adverse effects occur or may occur on the environment due to direct or indirect exposure to physical, chemical, or biological agents. This assessment is focused primarily on the interaction between populations, communities, and ecosystems [1, 10].

Human Health Risk Assessment (HRA) strictly examines the potential human health risks that may occur because of exposure to certain chemical, or biological substances and can result in short- or long-term consequences outside of the workplace [1]. In this case, the substances or materials - humans were exposed to, and the intensity or duration required to produce the adverse effects [3] are evaluated. Typically, health risks involve high probability, low consequence, and chronic exposure, making it difficult to determine cause and effect.

The flow diagram in (Fig. **2.3**) illustrates the two assessments. Regardless of the focus of the risk assessment study, it is divided into four steps.

```
                          ┌─────────────────────────────────┐
                          │  Environmental Risk Assessment  │
                          │                                 │
                          └─────────────────────────────────┘

┌────────────────────────────┐                            ┌────────────────────────────┐
│ Occupational health        │                            │ Wild life natural          │
│                            │                            │ vegetation; Agriculture    │
│ Environmental wellbeing    │   ┌────────────────────┐   │                            │
│                            │   │ Resources          │   │ Forests wetlands           │
│ Domestic livelihood        │   │ Air, water, soil, microbiota │                      │
│                            │   └────────────────────┘   │                            │
└────────────────────────────┘                            └────────────────────────────┘

                          ┌─────────────────────────────┐
                          │ Problem identification      │
                          └─────────────────────────────┘

┌────────────────────────────┐                    ┌────────────────────────────┐
│ Human health risk          │                    │ Ecological risk            │
│ assessment                 │                    │ assessment                 │
└────────────────────────────┘                    └────────────────────────────┘

┌────────────────────────────┐                    ┌────────────────────────────┐
│ Hazard identification      │                    │ Problem formulation        │
└────────────────────────────┘                    └────────────────────────────┘

┌────────────────────────────┐                    ┌────────────────────────────┐
│ Risk response              │                    │ Analysis                   │
└────────────────────────────┘                    └────────────────────────────┘

┌────────────────────────────┐                    ┌────────────────────────────┐
│ Risk characterization      │                    │ Risk characterization      │
└────────────────────────────┘                    └────────────────────────────┘
```

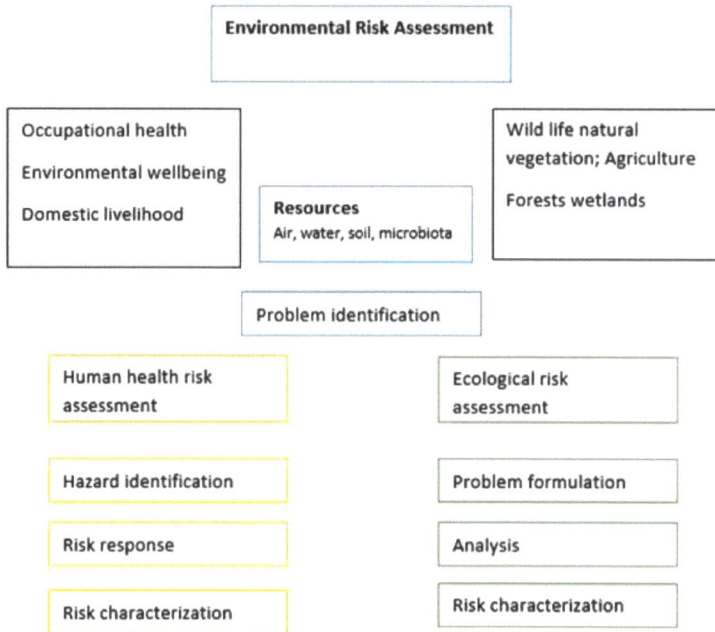

Fig. (2.3). Steps in the realization of human and ecological risk assessment [8].

The *Environmental Risk Assessment* begins with a clear identification of the problem/issue that may lead to a risk. During this phase, scientists, regulatory bodies, and managers participate in the assessment to define the goals and scope, reveal risk-related findings, and finalize the assessment timing. Following that, it is described in greater detail during the problem formulation stage.

PROBLEM FORMULATION STAGE/HAZARD IDENTIFICATION

Hazard identification or problem formulation defines the hazard and nature of the harm. The objective of this stage is to:

• Identify risk management goals and options;
• Identify natural resources of concern;
• Reach an agreement on the scope and complexity of the assessment [5].

The problem formulation phase must be clearly defined, including all constraints as well as subsequent risk management decisions and implementation. A clear and concise understanding of the problem will aid in understanding the level and type of risk assessment (quantitative/qualitative) to be used, as well as risk management decisions to be made.

The objectives of this stage compared to the (first) problem identification:

- Enhance the objectives of the risk assessment procedure;
- Determine which human/ecological entities are at risk;
- Determine the important characteristics that should be protected.

Hazard identification refers to the identification of the negative effects that a substance can have on human health, which requires the use of various toxicological and epidemiological data [5].

Dose-Response Assessment

A dose-response analysis is the second step in the risk assessment process (Fig. **2.4**).

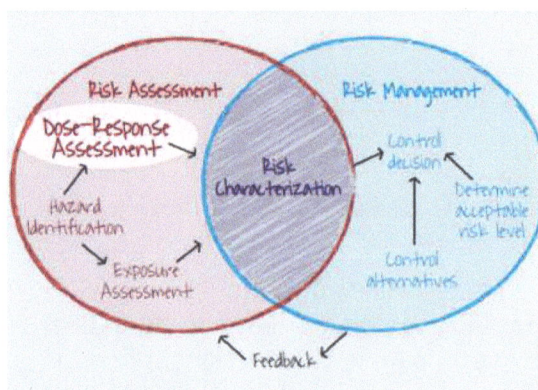

Fig. (2.4). Dose-response assessment within the risk assessment process (Image Source: ORAU, ©, https://www.toxmsdt.com/63-dose-response-assessment.html).

The dose-response relationship is established by assessing the relationship between the exposure and the adverse health effects, quantitating the hazards that were identified in the hazard identification step. In other words, during the dose-response assessment, the link between the dose and the adverse effects of the current hazard on the health of the population exposed to that risk is evaluated. The study's findings show the presence or absence of the dose-response effect, so the relationship between dose and toxic effect.

Chemical substances can have a variety of effects on humans and ecosystems. Specific information is gathered to assess the chemical agent's ability to cause quantitative toxicity. Most of the chemical toxicity information is derived from laboratory experiments in which the effects of the contaminant are observed after a specified exposure time. A dose-response relationship- a quantitative correlation that specifies the corresponding substance level of toxicity- is established based on these findings.

A dose-response assessment seeks to establish a mathematical relationship between the amount/concentration of contaminant to which a human is exposed and the risk of adverse consequences from that dose [8]. The results of experimental studies are illustrated in the figure below, with the abscissa representing the dose and the ordinate estimating the risk (Fig. **2.5**).

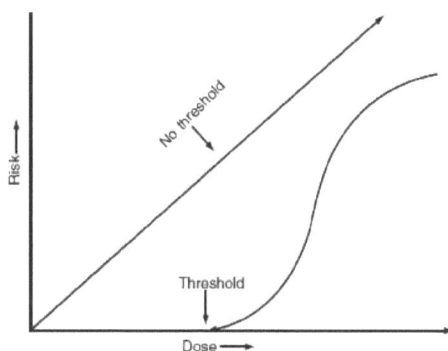

Fig. (2.5). Results obtained for dose-response assessment [9].

The Health risk assessment evaluates Carcinogenic or non-Carcinogenic health risks and is typically based on risk level estimation. Two representative lines summarize the plot from Fig. (**2.5**): no threshold and threshold. The threshold line for non-carcinogenic chemical responses represents the probability of adverse effects occurring below the line. Non-threshold is represented by carcinogenic doses and the possibility of additional cases per unit of time in a specified population (*e.g.,* 1×10^{-6} lifetime cancer risk).

US Environmental Protection Agency defines **Carcinogenic** or **Cancer Risks** (CR) as "*the incremental probability of an individual to develop cancer, over a lifetime, as a result of exposure to a potential carcinogen*" [9]. The Incremental Lifetime Cancer Risk (ILCR) is also defined as "*the incremental probability of a person developing any type of cancer over a lifetime as a result of a defined daily exposure to a given daily amount of a carcinogenic element over a 70-year lifetime*" [10]. ILCR is calculated by multiplying the **Slope Factor (SF)**, defined as the risk generated by a lifetime average amount of one mg/kg/day of carcinogen chemical (SF is contaminant specific and is a toxicity value that quantitatively defines the dose-response relationship), and the **Average Daily Dose (ADD)**, that represents the concentration of the chemical in milligrams per kilogram of body weight per day: $1/ (mg\ kg^{-1}\ day^{-1})$, through different exposure pathways. The equation defining ILCR is:

$$ILCR = SF \times AD \tag{2.2}$$

The likelihood that a receptor will develop non-cancer health effects because of long-term exposure to a specific chemical or chemical group is referred to as **non-Carcinogenic Risk**. The Dose-Response effects of non-Carcinogens allow for the existence of thresholds, or the amount of a substance or dose below which no toxic effect is observed (no observable toxic effect - NOAEL).

If a NOAEL is not available, a LOAEL may be used. LOAEL - lowest observed adverse effect level represents the lowest observed dose or concentration of a substance at which there is a detectable adverse health effect. An additional uncertainty factor is usually applied when a LOAEL is used instead of a NOAEL.

The reference dose (RfD) of a substance, which represents the intake or dose of the toxicant per unit body weight per day (mg kg^{-1} day^{-1}), is obtained by dividing the NOAEL (or LOAEL) by an uncertainty factor. The RfD is determined by using the following equation:

$$RfD = \frac{NOAEL}{VF1*VF2....*VFn} \tag{2.3}$$

The uncertainty factor is used to show the sensitivity differences between the population. RfD can be used to estimate non-cancer risk in the following way:

$$Risk = SF (CDI \times RfD) \tag{2.4}$$

Sf = slope factor

CDI = Chronic Daily Intake, represented by the average daily intake (mg day^{-1}) over the body weight (kg).

RfD is used in practice as a simple indicator of potential risk. Specifically, the Chronic Daily Intake (CDI) is regularly compared with the RfD indicator; if the CDI is lower than the RfD values, the risk is considered insignificant [9].

Exposure Assessment

According to the EPA's guidelines, exposure assessment is the process of estimating the magnitude, frequency, and duration of a stressor's exposure, as well as the number of people exposed, to estimate the concentration/doses of the stressor to which humans or the environment are exposed.

Environmental exposure can be estimated using direct or indirect methods. Direct exposure occurs when a substance is released into the environment (air, water, soil) from industrial facilities, for example. Thus, at the interface between the pollutant and the medium, the point of contact is calculated/measured. Indirect exposure can occur *via* the food chain or drinking water.

This stage is referred to as the analysis stage by different sources. In the case of Ecological Risk Assessment, this step evaluates the ecological responses to stressors under various exposure scenarios. As with Human Health Risk Assessment, this step identifies which receptors are exposed or are likely to be exposed, as well as the magnitude, frequency, and duration of exposure. During this stage, the pollutant/path stressors to the receptor is determined, as well as how exposure occurs [5].

Chemical substances can exist in the environment in a variety of forms after release:

- **Bioaccumulation** is the continuous accumulation of chemicals in a medium (pesticides, insecticides, *etc)*. This process occurs when an organism's capacity for absorption (skin, breathing) exceeds its capacity for elimination, and it is most common in animals [11];
- **Biomagnification** arises when contaminants (pesticides, metals) migrate into lakes, seas, and oceans and then enter the food chain [12];
- **Bioconcentration** is common in aquatic environments and refers to the chemical concentration of a contaminant that is greater than the water content in a living organism [11].

Researchers can estimate exposure by taking direct measurements of the amount of toxins in the environment (*via* environmental monitoring) or in the human body. Some countries have developed a computer-based program that employs predictive modeling techniques for determining the fate and transport modeling, as well as human exposure modeling. However, using indirect data to determine the exposure assessment is another option. Indirect method is used to determine data by combining predicted environmental concentrations of air, water, and soil with the consumption of water and food by the exposed groups [9].

Complete exposure assessment must include:

- Characterization of the physical settlement, climate, meteorological, geographical conditions, type of soil, groundwater, *etc*.;
- Characterization of the population that may be exposed;
- Identification of the exposure pathways (the source, the media that receives the

toxins), thus the data includes the fate and transport, physical and chemical characteristics of the stressor, and so on;
- Integration of the finding data in the exposure pathway model (source, fate and transport, exposure points and routes) [9].

Risk Characterization

For both health and ecological risk assessment, the last step of the risk assessment study is risk characterization. Throughout this phase, it the risk is estimated using the data from the previous phases, with a discussion of the overall degree of confidence in the risk values obtained.

Furthermore, in the case of Health and Ecological Risk Assessment, each risk assessment section must have its individual characterization. This is required for a clearer picture of the risk. Separate characterizations accompany the phases of hazard identification, dose-response and exposure assessment in Human Health Risk Assessment. Separate characterizations for Ecological Risk also accompany the analysis plan, the stressor-response profile, and the exposure profile. These key sections should conduct separate, component-by-component characterizations and include all findings, assumptions, strengths and limitations, and so on, resulting in the fundamental set of information conveyed in the Risk Characterization section (Table **2.3**) [5].

Table 2.3. Key points to be accomplished while completing a Risk Characterization phase [13].

Principle	Definition	Criteria for Good Risk Characterization
Transparency	Explicitness in the risk assessment process.	✓ Describe the assessment approach, assumptions, extrapolations, and use of models ✓ Describe plausible alternative assumptions ✓ Identify data gaps ✓ Distinguish science from policy ✓ Describe uncertainty ✓ Describe the relative strength of the assessment
Clarity	The assessment is free from obscure language and easy to understand	✓ Employ brevity ✓ Use plain English ✓ Avoid technical terms ✓ Use simple tables, graphics and equations

(Table 2.3) cont.....

Principle	Definition	Criteria for Good Risk Characterization
Consistency	The conclusions of the risk assessment are characterized in harmony with other EPA actions	✓ Follow statutes ✓ Follow agency ✓ Use agency information systems ✓ Place assessment in context with similar risks ✓ Define the level of effort ✓ Use review by peers
Reasonableness	The risk assessment is based on sound judgement	✓ Use review by peers ✓ Use the best available scientific information ✓ Use plausible information

Uncertainty in Risk Assessment

Uncertainty is a component of risk assessment evaluations. While estimating the risk, it is critical to understand the nature and magnitude of the uncertainty at the start of each risk assessment study. Examples of derived uncertainty include exposure assessment, analytical method limitations, and others.

Two approaches for characterizing uncertainty are commonly used in Risk Assessment evaluations. *Sensitivity analyses* and *Monte Carlo simulations* are commonly used for uncertainty characterisation. For **sensitivity analyses**, uncertain quantities of each parameter are varied (ex., average values), thus identifying how the changes affect the overall estimated risk. These analyses bring out a range of possible values of the risk and tell us which parameter is most important while establishing the size of the risk. However, for **Monte Carlo Simulation** all the given parameters are set out as random and uncertain. Considering this approach, a computer-based program is used to select the parameter distribution randomly and further solve the equations. This method is recurrent. The output results give us information of values for the exposure risk considering a specified probability, for example, at 55% or 80% [9].

Example of an Environmental Risk Assessment

Environmental risk assessment is being widely used around the globe as a low-cost mechanism to identify existing problems, incorporating chemical, biological, toxicological or ecological overviews. In Europe, there are a number of sites affected by the past activities; the same situation happened with the former ferrochromium works in Siechnice, Poland, used to manufacture different alloys. The site has been contaminated with heavy metals, especially chromium, in different chemical formulas. As time passed, the first step in solving the environmental issues was to perform an environmental risk assessment of the site.

In brief, chromium can be found in water, soil and air and be in different forms. There are 3 forms of Chromium: Cr (III), Cr (VI) and less common Cr (0), trivalent and hexavalent chromium are the most encountered, however, the effects differ significantly. Cr (III) stands for control of the lipid, peptide and glucose metabolism, on the other hand, Cr VI is a human carcinogen.

At present, the area poses a huge problem to the community nearby, as it is a source of uncontrolled release and spread in the soil, water and air, polluting wells, gardens and lowering living conditions.

The study was focused on both ecological and human perspectives, in both cases, the doses were determined in compliance with the assumed scenario and recommended facts by the US Environmental Protection Agency. The assessment was performed following the steps of exposure assessment, dose-response relation, risk characterization and uncertainty analysis, taking into consideration occupational, residential and children exposure. As final conclusions of the assessment showed, at first, a high concentration of both forms of chromium. For adults and children, the hazard is limited, however, for the occupational workers, results showed a high risk. The biggest hazard is associated with dust emissions to the environment. Herein protective measures should be applied to reduce workers' exposure to hazardous dust emissions [14]. As for environmental concerns in the investigated area and nearby, it was determined that Chromium hexavalent leads to root germination and growth reduction in the plants in other words, Cr affects biochemical and physiological processes in plants' productivity and yield [15]. As for animals, oral exposure to chromium leads to severe developmental and reproductive effects [16].

CONCLUDING REMARKS

The process of assessing environmental risk involves determining the likelihood or probability of a negative outcome or event resulting from pressures or changes in environmental conditions caused on by human activity. Based on this type of studies, government or private companies plan their next steps into reducing the negative effects and proceed on implementing *e.g.* various remedial actions, installing a newer technical equipment with proven less emissions. Results achieved within the Environmental Risk Assessment can lead even to closure of the facility if no best available technique (BAT) was identified. The basic steps of an ERA procedure have been described in the current chapter. In addition, it may be adapted or even developed based on the complexity of the site investigated.

REFERENCES

[1]   K. Flemström, R. Carlson, and M. Erixon, "Relationships between life cycle assessment and risk assessment", *Environmental Science,* 2004.

[2]   A.M.J. Ragas, "Trends and challenges in risk assessment of environmental contaminants", *J. Integr. Environ. Sci.,* vol. 8, no. 3, pp. 195-218, 2011.
[http://dx.doi.org/10.1080/1943815X.2011.597769]

[3]   A. Marta Schuhmacher, Mari Montserrat, Tsan Michael, and G. Sonnemann, *Integrated Life-Cycle and Risk Assessment for Industrial Processes and Products* 2nd. Taylor&Francis, 2018, p. 433.

[4]   C. D. M., and W. P. W. Robyn Fairman, *Environmental Risk Assessment: Approaches, Experiences and Information Sources* vol. 316. European Environment Agency, 2015, pp. 2014-2015.

[5]   E. P. A. EPA, "Conducting an ecological risk assessment", Available at: https://www.epa.gov/risk/conducting-ecological-risk-assessment (Accessed on: 2022).

[6]   TELLUS International, *Environmental risk assessment.* vol. 6. .

[7]   EUFIC, "Difference between hazard and risk", Available at: https://www.eufic.org/en/understanding-science/article/hazard-vs.-risk-infographic (Accessed on: 2022).

[8]   L. and P. Ministry of Environment, "Environmental risk assessment", In: *An approach for assessing and reporting environmental conditions.*, 2000.

[9]   I. L. Pepper, C. P. Gerba, and M. L. Brusseau, Environmental & Pollution Science vol. 53. Elsevier, no. 9, pp. 617-633, 2013.

[10]   D.M. Cocârță, A. Badea, M. Ragazzi, and T. Apostol, "Methodology for the human health risk assessment from the thermoelectric plants", *UPB Sci. Bull.,* vol. 70, no. 1, pp. 41-50, 2008.
[http://dx.doi.org/ISSN 1454-234x]

[11]   European Commision, "Aquatic Bioconcentration/Bioaccumulation", Available at: https://joint-research-centre.ec.europa.eu/eu-reference-laboratory-alternatives-animal-testing-eurl-ecvam/alterna-tive-methods-toxicity-testing/validated-test-methods-health-effects/aquatic-bioconcentrationbioac-cumulation_en (Accessed on: 2022).

[12]   K.G. Drouillard, "Biomagnification", In: *Elsevier, Encycl. Ecol.* vol. 1. Second Ed.. , no. 2, pp. 353-358, 2008.

[13]   EPA, Science Policy Council Handbook: Risk characterization Science Policy Council Handbook, p. 31, 2000.

[14]   A. Pawelczyk, F. Bozek, and M. Zuber, *Environmental Risk : Case Studies* University of Defence Brno: Czech Republic, 2018.

[15]   H. Oliveira, "Chromium as an environmental pollutant: Insights on induced plant toxicity", *J. Bot.,* vol. 8, p. 375843, 2012.
[http://dx.doi.org/10.1155/2012/375843]

[16]   "EPA Chromium Compunds", Available at: https://www.epa.gov/sites/default/files/2016-09/docu-ments/chromium-compounds.pdf

CHAPTER 3

Risk Management and Principles

R. Lupu[1,*]

[1] *University POLITEHNICA of Bucharest, Faculty of Energy Engineering, Splaiul Independentei 313, RO-060042 Bucharest, Romania*

Abstract: The chapter discusses the concept of risks in the economy and how it affects the everyday operations of a company. Risks arising can result in economic losses, damage to facilities and equipment, and, most importantly, workplace accidents. Risk management nowadays has developed in both concept and practice and has become an industry in countries with a functional market economy. The chapter emphasizes that risk management is a crucial aspect of global management that requires information from various fields, such as economic, technical, legal, statistical, and psychological, to maintain the risk at the minimum level. Moreover, effective risk management can lead to sustainable development for humans, the environment, and businesses.

Keywords: Risk Assessment, Risk Management, Risk Control, hazard, Risk evaluation, Risk control, Risk-based management, Risk monitoring, Risk communication.

INTRODUCTION

Facing various situations involving risks, as well as being exposed to hazards, represents daily life for humans in both situations, at work and at home. At the European level, to ensure a healthy and safe workplace, it is an important factor within a fully completed Risk Management.

Environmental Risk Assessment includes two distinct components: Risk Assessment and Risk Management. Risk Assessment, as defined in previous chapters, is the process of assessing the risk posed by substances released into the environment because of anthropic activities, as well as the identification of affected receptors (local, community, and so on), and the assessed risk level.

Risk Assessment may estimate the expected rate of illness among the vulnerable population (population at risk). In the case of Risk Management, however, the results of the Risk Assessment are integrated into various conditions and moni-

---

* **Corresponding author R. Lupu:** University POLITEHNICA of Bucharest, Faculty of Energy Engineering, Splaiul Independentei 313, RO-060042 Bucharest, Romania; E-mail: lupurusalina@gmail.com

tored from an economic or legal standpoint. As a result, risk reduction decisions are made to reduce or eliminate the risk completely. Risk Management is the process of deciding what to do about a hazard, the population at risk, or adverse effects, putting the decision into action, and evaluating the results. It also refers to program or authority-level decision making, such as determining which hazards should be managed and in what order. Risk Management can benefit from comparative (or relative) risk analysis and cost-benefit analysis [1].

Some examples of Risk-based Management actions are:

- Decide the amount of a substance to be discharged in the natural habitat;
- Decide which substances are safe to be stored at the hazardous waste disposal facility;
- Decide what strategy to be used for a hazardous contaminated site: actions for cleanup or remediation;
- Set permit levels for waste discharge, storage, and transport;
- Establish national ambient air/noise quality standards and thresholds for the water contamination.

More specific information for a better understanding is presented in (Fig. **3.1**).

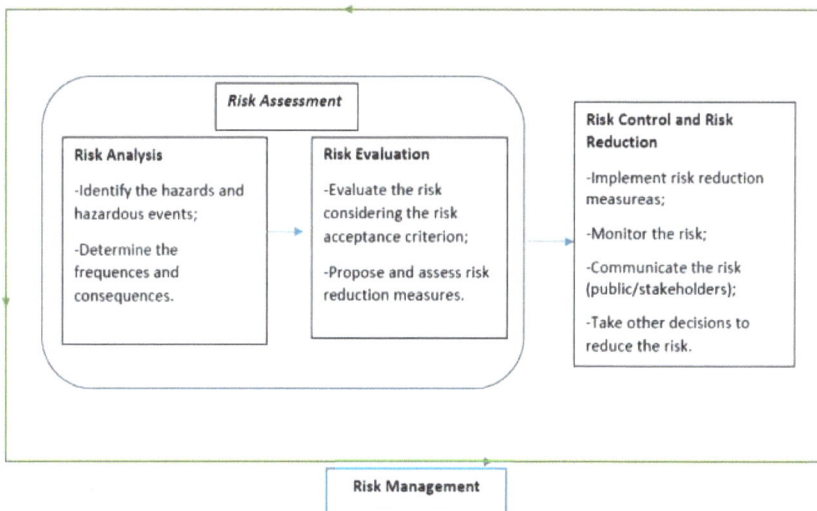

Fig. (3.1). Main elements of Risk Management process (Source [2]).

Risk management is defined as the management process that aims to analyze and assess potential hazards and propose effective risk control measures to reduce or eliminate any potential harm to people or the environment.

Risk Management, in relation to Risk Assessment, is composed of four stages process. The decision-making must be an iterative process. The information gathered between the stages will determine to go back and to revise the scope and principles. (Fig. **3.2**) illustrates the Risk Management stages.

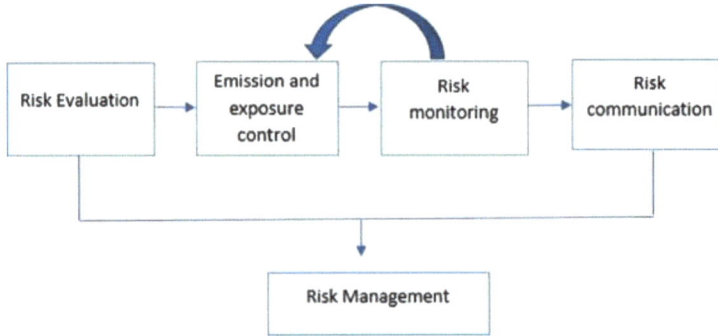

Fig. (3.2). Risk Management - main phases (Source [2]).

REQUIREMENTS OF RISK MANAGEMENT PROGRAM

The Risk Management Program is the formal process used to quantify, classify, and mitigate risks that Environmental Risk Professionals may discover or define in the context of environmental issues. Risk Management is the final process in the Environmental Risk Assessment and must be completed before the decision to mitigate/eliminate the Hazard/Risk can be implemented. In the risk management stage, there are several steps that must be taken into consideration for a positive impact on decision-making. Different phases in the Risk Management stage should be followed in a cycle mode (as shown in Fig. (**3.3**)) to ensure that the established decisions have a positive impact and that continuous improvement occurs throughout the process [3].

Fig. (3.3). Risk management cycle mode (Source [4]).

To **Identify** the hazard, it will be considered what hazards or risks are present onsite that could cause damage or harm to workers/environment/nearby inhabitants.

To **Assess** risk, the likelihood and consequence diagram will be used to determine the level and severity of the risk/hazard.

To **Implement** control/**Treat** the risk, it will be determined what measures or actions are appropriate and available to eliminate or reduce the risk/hazard.

To **Check** controls or **Monitor** and **Report**, the measures/actions implemented will be reviewed to ensure they are effective and efficient [3].

The probability and consequence diagram aids in defining the risk and the exposed receptors. The likelihood of receptor exposure as well as hazard/risk occurring could be **High**, **Medium**, **Low**, or **Very Low**. As shown in (Fig. **3.4**), the interpretation of the risk results is given by combining the probability with the potential consequence of the hazard. The diagram will be discussed in detail in the Risk Assessment step.

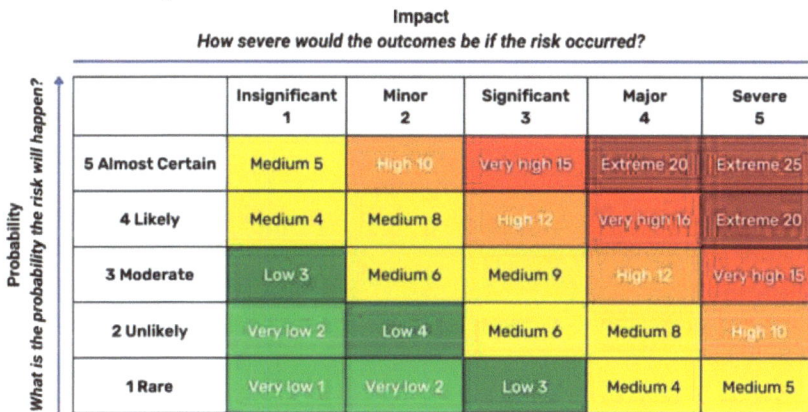

Impact
How severe would the outcomes be if the risk occurred?

		Insignificant 1	Minor 2	Significant 3	Major 4	Severe 5
	5 Almost Certain	Medium 5	High 10	Very high 15	Extreme 20	Extreme 25
	4 Likely	Medium 4	Medium 8	High 12	Very high 16	Extreme 20
	3 Moderate	Low 3	Medium 6	Medium 9	High 12	Very high 15
	2 Unlikely	Very low 2	Low 4	Medium 6	Medium 8	High 10
	1 Rare	Very low 1	Very low 2	Low 3	Medium 4	Medium 5

Probability — What is the probability the risk will happen?

Fig. (3.4). The probability and consequence diagram (Source [5]).

RISK MANAGEMENT: IDENTIFY HAZARDS, ASSESS RISKS, IMPLEMENT AND CHECK CONTROLS

Identification of the Risk

The hazards associated with the activities that the facility/company is performing, everything that can cause harm to people and the environment, are established during the Identification of Risk phase. Table **3.1** contains a list of hazards and common sources of pollution.

Table 3.1. Example list of hazards present in a facility [2].

Hazard	Description	Sources of Pollution
Emissions from explosions or fire	Dangerous situation for human life, environment and facility, explosions and fire present a hazard derived from toxic smoke, spreading dust.	- Hot surfaces - Poor storage - Smoking in the facility - Dust - Electrical hazard
Chemical Spills	Release of chemical substances can contaminate soil and water, which is a threat to human health.	- Leaking containers - Poor storage of chemicals and handling
Air contaminants	Toxic substances being released into the atmosphere from the manufacturing processes activities are considered major pollutants.	- Boilers - Furnaces - Uncovered solvents - Vents

There are various methods of identifying hazards; typically, the company/facility lists the standard methods of inspection and identification of non-conformities in the management guide. Commonly, once a hazard has been identified, it is recorded in a specific register. The best way to identify the hazard is to conduct an onsite inspection at the facility, which should include not only an inspection of the building, equipment, or structures, but also a check-out of the operating procedures, systems, and so on. The auditor/inspector should always consider whether there are any nearby receptors that may be harmed, such as houses, waterways, forests, and so on, as well as the exposure pathways. Meetings with employees and stakeholders are a good way to identify hazards because they involve multiple parties in the assessment. Furthermore, available site history information is a good source of identification of whether there was a hazard, if that hazard persists and may have a potential impact [3].

Risk Assessment

Once the hazards/risks are identified, they must be evaluated further to determine how they may harm the environment/population, the severity of that harm, and the likelihood of the hazard occurring.

In the beginning, to assess the likelihood of a hazard causing incidents or harm, the following factors must be considered:

• Determination of the likelihood that the hazard/contaminant will cause harm/damage;
• Evaluation of the hazard's consequences;

• Calculation of the hazard's risk.

There are several questions to ask whenever assessing the consequence of the hazard:

1. What kinds of harm could be caused by a/the hazard?

2. What factors might have an impact on the severity of the damage?

3. How might human health or the environment be harmed or damaged?

The probability and consequence diagrams are widely used to calculate the risk rating after receiving all the information.

Implement Controls

After evaluating each hazard, the auditor or inspector implements one or more controls that either remove the threat or lessen the likelihood and/or severity of a hazardous incident.

Controls can take many forms, but the three most common are as follows:

   I. Elimination/Avoidance Controls
  II. Engineering/Physical Controls
 III. Educational (Awareness) Controls

Using the hierarchy of controls diagram, risk must be prioritized from highest to lowest level, and actions must be taken in accordance with each step. The control hierarchy provides us with potential risk management approaches and prioritizes them in terms of the order that should be considered, ensuring firstly the most effective approaches (which is eliminate hazard) and the least effective (which is administrative controls) as the last option.

Elimination of the hazard should be attempted first at the top of the control hierarchy. We want to physically remove the hazard wherever possible and then try a substitution process, which means replacing one hazard with one that is slightly less harmful, or possibly significantly less harmful, or engineering controls that are embedded together and ensure isolating people from the hazard. Administrative controls, which may include changing people's behavior at work (Fig. **3.5**), are between the last options [3].

Having followed the diagram of the hierarchy of controls, the hazard is controlled by considering the information from the previous steps and applying it from the top of the hierarchy to the bottom.

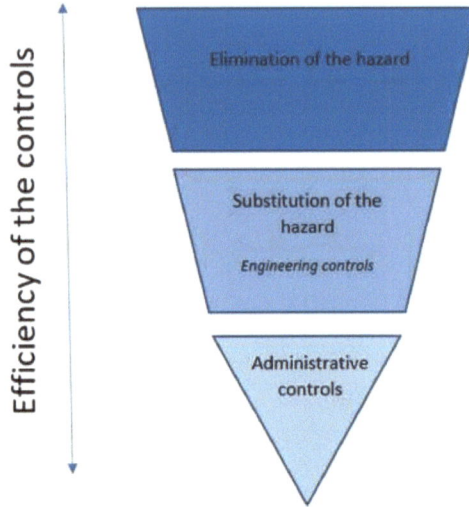

Fig. (3.5). Hierarchy of controls applied in risk management [3].

Check Controls/Monitor and Report

As a direct supervisor, we will closely monitor the efficiency of the decisions regarding the elimination or reduction of the hazard in the final step of the Risk Management and re-do practically the same steps as in the first (identification of the hazards). Risk monitoring is essential in Environmental and Human Risk Management because it ensures that risk mitigation or reduction is effective.

Some conditions are monitored on a regular basis to ensure effective risk control:

• Site inspections and audits on a regular basis;
• Interacting with employees, contractors, tenants, and property owners;
• Inspecting, testing, and maintaining risk management systems;
• Utilizing readily available data, such as manufacturer/supplier instructions;
• Examining records and data such as incident and near-miss reports [2, 6].

CONCLUDING REMARKS

Risk assessment procedure is an important step of the Environmental Risk Assessment study, in order to achieve the efficiency of the steps implemented for environmental damage reduction. This step ensures that the government or the private company has a tight understanding and manage on the steps implemented for eliminating or reducing the threat identified. Thus, it is highly recommended to invest time, research and dedicated personnel for going through this procedure – be cautious on the activity implemented and possible overview on the outcome. Delivering a high-quality work on reducing or eliminating the damage, will lead to improvement on human life quality and reduce animal/birds/ insects extinction as proven by recent studies.

REFERENCES

[1]   I.V. Muralikrishna, and V. Manickam, "The realities of risk-cost-benefit analysis : Risk assessment", In: *Science* vol. 80. Environmental Management, p. 350, 2015.

[2]   I.V. Muralikrishna, and V. Manickam, *Environmental management: science and engineering for industry*. Butterworth-Heinemann publications, p. 664, 2017.

[3]   E.P.A. Victoria, *Assessing and controlling risk: A guide for business Assessing and controlling risk for business 2* EPA victoria, pp. 1-16, 2018.

[4]   MI-GSO, "The risk management process: 4 essential steps", Available at: https://www.migso-pcubed.com/blog/pmo-project-delivery/four-step-risk-management-process/ (Accessed on:20-Au--2022).

[5]   Safety culture, "A guide to understanding 5x5 risk matrix", Available at: https://safetyculture.com/topics/risk-assessment/5x5-risk-matrix/ (Accessed on:14-Aug-2022).

[6]   SIM E.S., and Viorel D., *Environmental risk management and sustainable enterprise.*, pp. 435-444, 2011.

CHAPTER 4

Risk-Based Approach for Contaminated Soil Management

C. Streche[1], M.C. López-Escalante[2], F.P. Martín Jiménez[2] and Diana Mariana Cocârță[1,*]

[1] *University POLITEHNICA of Bucharest, Faculty of Energy Engineering, Splaiul Independentei 313, RO-060042 Bucharest, Romania*

[2] *Department of Chemical Engineering, University of Malaga, Faculty of Science, Malaga, Spain*

Abstract: The current chapter illustrates aspects of sustainable soil management, the basic concepts for site investigation based on the calculation of the risk associated with the various chemicals that affect the structure, quality, and functions of the soil, as well as the highlighting of the most important remedial strategies used to reduce pollution. As a result, the soil can be used for a variety of industrial and civil purposes. Sustainable soil management is a concept based on technical-scientific and economic knowledge. Simultaneously, policy actions are taken to maintain and increase soil productivity, protect biodiversity, reduce risk, and protect natural resource potential by preventing soil quality degradation and supporting ecosystem services. In the decision-making process for Assessing Human Health Risk for contaminated sites, the development of the site's conceptual model is recommended for a better understanding of the evolution of the respective site's situation. The development of a Conceptual Site Model (CSM) is useful for assessing the contamination risks of any site because it identifies the sources of hazards, potential receptors (people, ecology, and infrastructure), and exposure pathways.

Keywords: Soil pollution, Risk assesment, Risk-Based Land Management (RBLM), Heavy metals, Conceptual Site Model (CSM), Remediation technologies, Electrochemical remediation.

INTRODUCTION

Due to pedogenic factors like climate, microorganisms, vegetation, and landforms, the soil has developed from rocks. The rocks have undergone significant transformations over time so that the soil appears to be a natural body distinct from the parent rock. The regeneration time is long, so it takes between 300 and 1000 years to form 3 cm of soil naturally, and 70000 years to form

---

* **Corresponding author Diana Mariana Cocârță:** University POLITEHNICA of Bucharest, Faculty of Energy Engineering, Splaiul Independentei 313, RO-060042 Bucharest, Romania; E-mail: dianacocarta13@yahoo.com

Diana Mariana Cocârță (Ed.)
All rights reserved-© 2023 Bentham Science Publishers

20 cm. [1] Soil is the environment on the earth's crust's surface where human life occurs. It is an essential and extremely complex resource composed of mineral particles, organic matter, water, air, and living organisms. Soil is a dynamic system that has many functions and is essential for human activities and ecosystem survival.

Pollution is a consequence of human activity, particularly social and economic activity. From a historical standpoint, environmental pollution appeared at the same time as man, but it has evolved and diversified in response to the evolution of human society, becoming one of the most pressing concerns of scientists and technologists, states and governments, and the entire global population today. This is because the threat posed by pollution has grown and continues to grow, requiring urgent national and international action in the spirit of pollution-fighting ideas.

According to the European Environment Agency (EEA), statistics on the number of decontaminated sites and the number of contaminated and potentially contaminated sites were updated at the European level in 2017 (Fig. **4.1**).

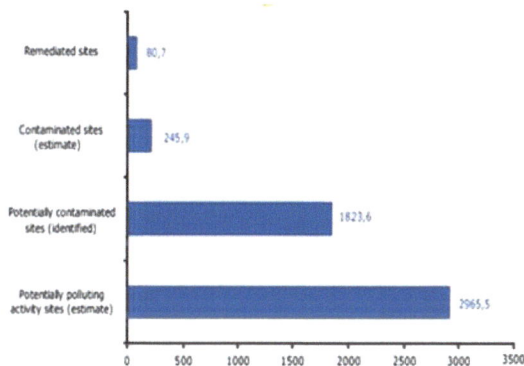

Fig. (4.1). Estimated situation of polluted sites in 2017 in EU countries (number of sites x 1000) [2].

The main economic-industrial activities that caused soil pollution are also represented as a percentage of the number of sites where preliminary investigations were completed, with 2012 as the reference year (Fig. **4.2**).

Prior to the appearance of this concerning context, the general belief was that the soil has an almost limitless capacity for self-purification and resilience. This fact was proven to be false because only the instantaneous reactions of the soil to the various disturbing polluting actions to which it is subjected were quantified. In such cases, nature lacks the ability to correct the negative effects produced in a relatively short period of time, necessitating intervention to limit and eliminate polluting sources of any kind.

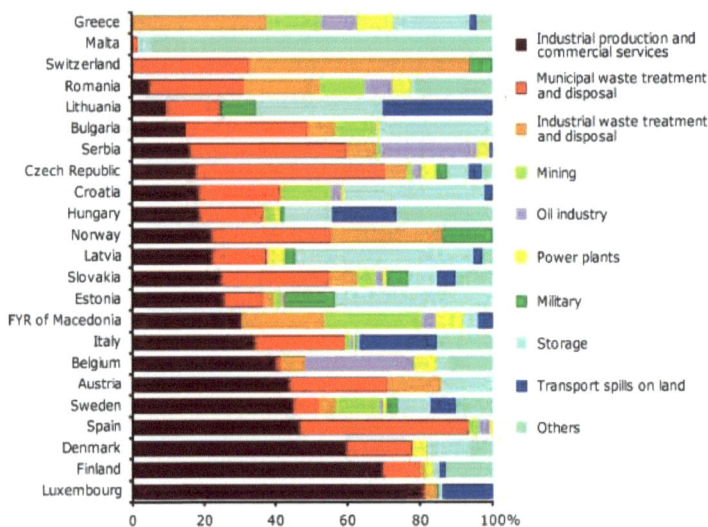

Fig. (4.2). The main industrial and economic sources responsible for environmental and human health pollution [3].

When viewed through the lens of harmful effects on human health, environmental pollution has manifested itself in various types of pollution, as follows:

o **Biological pollution**, the oldest and most well-known type of pollution, is caused by the elimination and spread of microbial germs, mostly pathogens, in the environment by humans and animals. The primary threat posed by biological pollution (bacteriological, viral, and parasitological) is the spread of epidemics.

o **Chemical pollution** is the elimination and spread of various chemical substances in the environment. Chemical pollution is becoming increasingly visible, both by increasing the level of pollution and, more importantly, by diversifying it. The main danger of chemical pollution is the highly toxic potential of these substances, which causes problems with both normal plant growth and human and animal health.

o **Physical pollution** consists primarily of soil erosion caused by environmental factors (the action of wind and rain) and desertification resulting from climatic variations and human impact. Acoustic noises and noise pollution have a significant impact on human societies. The human body was created to tolerate a certain level of noise and vibration; exceeding this threshold can cause illness, embarrassment, or even disruption. Thermal pollution is perhaps the most recent form of physical pollution, with significant effects on the environment, particularly on water and air, as well as indirectly on population health. The cooling systems of conventional and nuclear power plants, the chemical and metallurgical industries, and naval transport are the primary sources of thermal pollution [4].

Physical pollution is less well-known than biological and chemical pollution due to its wide variety and recent emergence, necessitating special investigation and research efforts to be mastered in the future.

SOIL POLLUTION AND HEALTH

Soil has a wide range of effects on human health. There is a direct link between soil health, pollution level, and human health. As a result, the impact of the polluting environment on the human body is diverse and complex. It can range from minor inconveniences in human activity, known as discomfort, to severe health problems.

From a historic perspective, the first observations and research on the impact of environmental pollution on human health were made following the appearance of acute effects caused by particularly high concentrations of pollutants in the environment, which had severe implications for the human body.

Polluting substances have a wide range of effects on the human body. These are directly affecting internal organs and causing functional instabilities at the level of human systems, with the effects amplified by their cumulative nature inside the human body [5]. (Fig. **4.3**) presents the main effects of soil contaminants on human health, indicating the organs or systems affected as well as the contaminants responsible for them.

Heavy metals, organic pollutants, and radioactive substances are the three main types of substances that can harm human health.

Heavy Metals

Heavy Metals can be found in the atmosphere, the lithosphere, the hydrosphere, and the biosphere. Heavy metals extracted from the earth's crust are used for a variety of industrial and economic purposes. Some of these heavy metals have direct or indirect effects on the human body. Other heavy metals, such as copper, cobalt, iron, nickel, magnesium, molybdenum, chromium, selenium, manganese, and zinc, play important physiological and biochemical roles in the body. On the other hand, different heavy metals are toxic to the body in high doses, while others, such as cadmium, mercury, lead, chromium, silver, and arsenic, have delirious effects in the body, causing acute and chronic toxicities in humans [7].

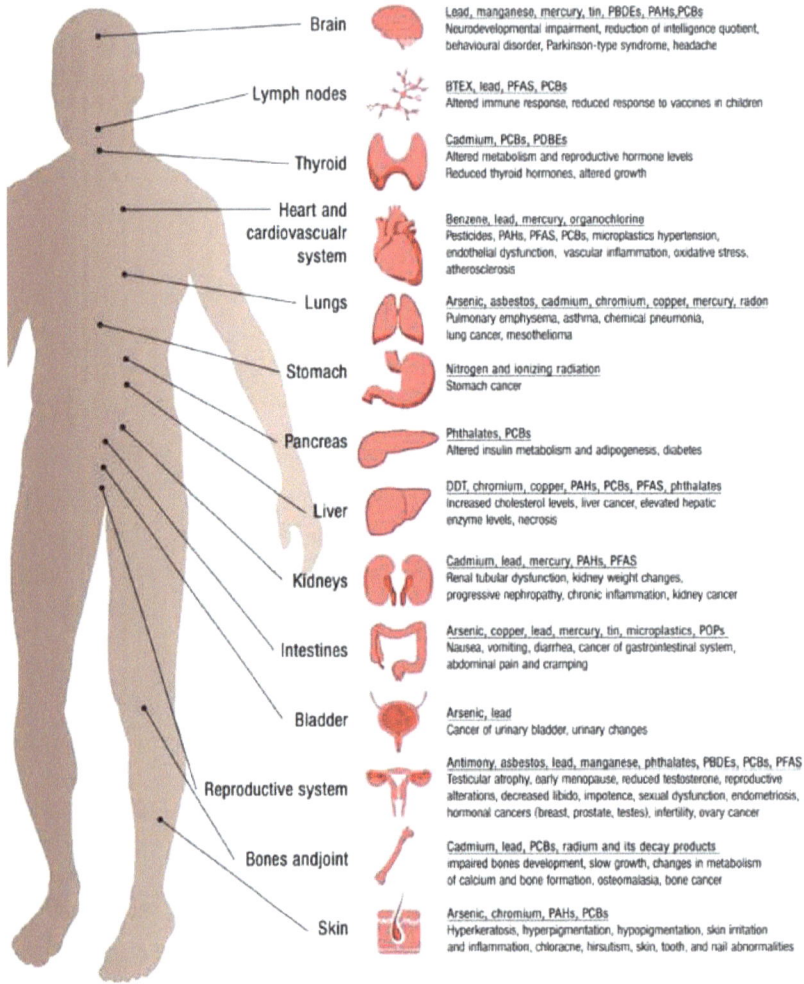

Fig. (4.3). Main effects of soil contaminants on human health, indicating the organs or systems affected and the contaminants causing them. (PCBs, polychlorinated biphenyls; PBDEs, polybrominated diphenyl ethers; PFAS, per- and polyfluoroalkyl substances; POPs, persistent organic pollutants; BTEX, refers to the chemicals benzene, toluene, ethylbenzene, and xylene) [6].

Heavy metals can harm and degrade organs such as the brain, kidneys, lungs, liver, and blood. Acute or chronic heavy metal toxicity can occur. Long-term heavy metal exposure can cause muscular, physical, and neurological degeneration, similar to diseases like Parkinson's disease, multiple sclerosis, muscular dystrophy, and Alzheimer's disease. In addition, long-term chronic exposure to heavy metals can lead to cancer [8].

Some heavy metals' various health effects are highlighted next:

• Arsenic

Arsenic toxicity can occur either acutely or chronically. Acute arsenic poisoning can cause blood vessel and gastrointestinal tissue destruction, as well as heart and brain damage. Chronic arsenic toxicity, also known as arsenicosis, is typically characterized by skin manifestations such as pigmentation and keratosis [9].

• Lead

Lead poisoning is toxicity caused by lead exposure. In both children and adults, lead poisoning primarily affects the gastrointestinal tract and the central nervous system [10]. Acute or chronic lead poisoning can occur. Chronic lead exposure can cause birth defects, mental retardation, autism, psychosis, allergies, paralysis, weight loss, dyslexia, hyperactivity, muscle weakness, kidney damage, brain damage, coma, and sometimes even death [9].

• Mercury

The elements mercury and oxygen combine readily to form both inorganic and organic mercury. The kidneys, brain, and developing fetus can be harmed by long-term exposure to high concentrations of metallic, inorganic, and organic mercury [11]. Because organic mercury has a lipophilic nature, it can easily pass-through cell membranes. Increased exposure to mercury can alter brain functions and cause tremors, shyness, irritability, memory issues, changes in hearing or vision, and other symptoms because mercury and its compounds have an adverse effect on the nervous system. While short-term exposure to organic mercury poisoning can result in depression, tremors, headaches, fatigue, memory issues, hair loss, *etc.*, short-term exposure to metallic mercury vapor can cause vomiting, nausea, rashes, diarrhea, lung damage, high blood pressure, *etc.* The diagnosis of mercury poisoning in such cases can be challenging because these symptoms are also typical of other diseases or conditions [9].

• Cadmium

The effects of cadmium and its compounds on human health are numerous. The inability of the human body to excrete cadmium exacerbates the negative health effects of cadmium exposure. In actuality, the kidneys re-absorb cadmium, limiting its excretion. Ingestion of higher doses of cadmium can cause gastric irritation, which can result in vomiting and diarrhea, while short-term inhalation exposure can cause severe lung damage and respiratory tract irritation. Cadmium accumulates in the bones and lungs as a result of long-term exposure. Cadmium exposure can harm bones and the lungs [12]. Animal and human studies have shown that cadmium can cause osteoporosis (damage to the skeleton), which can lead to bone mineralization.

• Chromium

The most toxic form of chromium is hexavalent, but some other forms, like Chromium (III) compounds, are much less dangerous to human health and rarely result in any issues. Chromite (VI) can be corrosive to the body and trigger allergic reactions. As a result, inhaling large amounts of chromium (VI) can irritate the nasal mucosa and result in nasal ulcers. Additionally, it can harm sperm and the male reproductive system, as well as cause anemia, irritation and ulcers in the small intestine and stomach. Chromium allergy symptoms include extreme skin redness and swelling. Chromium (VI) compound exposure in humans can have deadly consequences for the cardiovascular, respiratory, hematological, gastrointestinal, renal, hepatic, and neurological systems [13].

• Iron

When exposed topically, orally, or inhaled, iron salts such as ferrous sulfate, ferrous sulfate heptahydrate, and ferrous sulfate monohydrate have low acute toxicity. Other forms of iron, however, are extremely harmful to your health. Four stages of iron toxicity can be seen. Gastrointestinal symptoms like vomiting, diarrhea, and gastrointestinal bleeding characterize the initial stage, which begins six hours after an iron overdose. Six to twenty-four hours after the overdose, there is a progression to the second stage, which is regarded as a latent period of apparent medical recovery. Between 12 and 96 hours after the onset of clinical symptoms, the third stage sets in and is marked by hypotension, shock, lethargy, liver necrosis, tachycardia, and metabolic acidosis, and can occasionally result in death [14].

After the iron overdose, the fourth and final stage typically starts 2–6 weeks later. Gastrointestinal ulcers and the emergence of strictures characterize this stage. Since excessive iron consumption raises the risk of cancer, meat-consuming countries are at risk of developing the disease.

• Manganese

Although manganese is a necessary metal for the body, it recently gained attention due to the introduction of the gasoline additive methylcyclopentadienyl manganese tricarbonyl (MMT), which was known to be toxic. The development of a Parkinson's disease-like, including tremors, gait disturbances, postural instability, and cognitive impairment, has been linked to MMT and has been called an occupational hazard [15].

The neurotoxic effects of high levels of manganese exposure can occur. The neurological condition known as manganism, which is brought on by manganese, is characterized by rigidity, action tremor, a mask-like expression, gait disturbance, bradykinesia, micrographia, memory and cognitive dysfunction, and mood disturbance [16]. Manganism has many of the same symptoms as Parkinson's disease.

Persistent Organic Pollutants (POPs)

POPs are among the substances that are most dangerous to human health because of their extremely dangerous poisonous properties. Furthermorer, they demonstrate a high level of resistance to degradation as well as an accumulation characteristic in both the environment and living organisms. POPs can be easily deposed far from the source of emission and transported through the atmosphere over long distances.

The following categories can be used to classify the top twelve toxic persistent organic pollutants that have a negative impact on both the environment and human health:

o **Pesticides:** aldrin, dieldrin, dichloro-diphenyl-tetrachloroethane (DDT), heptachlor, mirex, chlordane, taxofen, endrin

o **Industrial chemicals:** hexachlorobenzene (HCB), polychlorinated biphenyl (PCB)

o **Secondary combustion products:** dioxins, furans [17].

POPs act on the human body in high concentrations to increase the risk of cancer, cause anomalous development, reduce fertility, reduce immunity, and inhibit cognitive function. The effects of these substances on the embryo, fetus, and young children are particularly serious [18].

The prevalence of breast cancer and human exposure to DDT are directly related, according to research. Because DDE, a metabolite of DDT, was found in the blood serum of sick women and people who had contact with DDT, the mammary gland can be thought of as a target organ. Aldrin, Dieldrin, Heptachlor, Chlordane, Endosulfan, and other chlorinated cyclodienes have been shown to have more pronounced toxic properties than DDT. Contrarily, the toxic effects of chlorinated cyclodienes manifest as convulsions, drowsiness, persistent spasms, headaches, *etc.* Most of these POPs are cancer-causing and frequently result in liver tumors. The body of a woman experiences the toxic effects of pesticides. Epidemiological studies have shown that the frequency of pathological conditions and functional disorders of the reproductive system in women also rises as pesticide application is intensified. Similar effects were also observed regarding male reproduction, as evidenced by the recent growth in cases of male sterility [19].

The most common pesticides with damaging consequences on both human health and the environment are:

• DDT (dichloro-diphenyl-trichloroethane)

DDT is a pesticide that has been extensively used on agricultural crops, particularly cotton, to control diseases like malaria and typhus that are spread by insects acting as their vectors (mosquitoes, fleas, lice). In many nations, use of the product was either prohibited or severely restricted.

Breast cancer risk and chronic disease are both linked to long-term DDT exposure. DDT exposure can also result from consuming specific foods, as the pesticide has been found in meat, eggs, vegetables, fruits, and even infants' milk.

• Heptachlor

Heptachlor is a substance that has toxic and dangerous effects on the human body. It is mainly found in adipose tissues and is absorbed through ingestion and dermal contact. The immune system is impacted by heptachlor, which has been identified as a possible carcinogen.

In addition to being quickly fixed in aquatic sediments and bioconcentrating in living organisms' fats, it is volatile enough to be dispersed in the atmosphere.

• Aldrin

Aldrin is a highly toxic insecticide that may be harmful to humans. The lethal dose for adults is thought to be 5g, or 83 mg/kg body weight. The symptoms of poisoning with this insecticide include headaches, nausea, vomiting, muscle contractions, spasms, and convulsions. Exposure to this insecticide can occur through the daily consumption of certain products (such as dairy products). Aldrin has a fatal effect on humans, birds, and fish, and dieldrin residues have also been found in infants' breast milk. Up to 97% of dieldrin in the environment comes from aldrin, used as an insecticide.

• Dieldrin

Because of its poor solubility in water, high stability, and semi-volatility, dieldrin has favorable physical and chemical characteristics that facilitate long-distance transportation. As a result, it was found in both the environment and Arctic organisms.

Because of the long half-life (4–7 years) of aldrin and dieldrin used in agriculture, residues can be present for a very long time. Aldrin and dieldrin remnants have been found in dead animals, eggs, predators, fish, amphibians, invertebrates, and soil.

• Endrin

Endrin is a foliar insecticide used mainly for large crops (cereals, cotton, *etc.*), but it is also used in the fight against rodents (rodenticide). Due to its toxicity, it was banned in many countries. Endrin is very dangerous for the human body by acting on the nervous system. In adults, the lethal dose is estimated at 10 mg/kg body weight. The effects of endrin on soil and fungi are minimal, but it is very toxic to fish and aquatic invertebrates, being rapidly metabolized by animals. It reaches the atmosphere through evaporation, being able to contaminate surface waters through rain. Its half-life is long, persisting in the soil for 12 years.

• Chlordane

Chlordane is used on vegetables (potatoes), cereals (corn), technical plants (beet and sugar cane, oilseeds, jute, cotton), and fruit trees. It is a broad-spectrum insecticide used on agricultural crops. Although chlordane was frequently applied to combat termites, it is currently prohibited in several countries. Chlordane is a highly toxic substance that is harmful to the human body and can be consumed in certain foods. The lethal dose for adults is thought to be 25–50 mg/kg body weight.

• Hexachlorobenzene

Hexachlorobenzene is toxic and dangerous, and the lethal dose for adults is estimated to be 0.13 mg/kg body weight. The signs and symptoms of HCB ingestion include dermatosis, colic, hyperpigmentation, severe weakness, and debility. Infant mortality can increase to 95% due to the compound's harmful effects on reproduction and the genital system. It crosses the placenta from mother to fetus and is present in breast milk.

Due to the mobility and chemical stability of hexachlorobenzene, which is widely distributed in the environment, it has been found in air, water, sediments, soil, and living things all over the world. It is extremely toxic to aquatic life.

• Dioxins and Furans

Herbicide production, industrial accidents, burning of chemical substances, and uncontrolled waste burning are all ways that people can be exposed to dioxins and furans. Dioxins and furans have been identified as potential human carcinogens.

Dioxins and furans are occurring mixtures of 210 substances, 17 of which are highly toxic. Seveso dioxin, one of these, is thought to be the most toxic substance ever created by humans. Dioxins are released into the environment when pesticides and other chlorinated substances are used.

Dioxins and furans are very stable and persistent in the environment, and they are only slightly soluble in water. Even 10–12 years after the initial exposure, these substances were found in the soil.

Physical characteristics that favor their transport over long distances include high stability, poor solubility in water, and semi-volatility. Dioxin exposure in animals results in lower fertility, genetic disorders, and embryo mortality [20].

Radioactive Substances

The unintentional and unwanted deposition of a radioactive substance on various surfaces or in a solid, liquid, or gas (including the human body) is known as radioactive contamination (or radiological contamination) [21]. The most harmful type of soil contamination is radioactive pollution, the effects of which can last for decades or even centuries. Radionuclides are substances that can be found in soil naturally or because of human activity (such as nuclear and medical waste) and are associated with illnesses like cancer and leukemia [22].

Radiation can ionize molecules, break chemical bonds, or produce heat when it interacts with living cells. The most severe biological harm is caused when molecules are broken up or ionized by radioactive emissions. By eradicating cellular DNA, radiation can have an impact on biological systems. Cancer can develop if cells divide too rapidly due to improper DNA repair.

Following some unfortunate events, such as the nuclear power plant disasters at Chernobyl (on April 26, 1986) and Fukushima (on March 11, 2011), the world's population has once again become aware of the grave dangers posed by radioactive contamination and its effects on human health. According to a recent study by Komissarova and Paramonova in a region still exceeding the safety standard for the isotope 137Cs by 3.5 to 6 times in 2017 [23], some areas were still affected by the Chernobyl accident. The radioactive isotope Cs-137, which is expected to have a long-term impact on agricultural products and human health, was thought to have contaminated the soils close to the nuclear power plant after the Fukushima accident [24].

RISK-BASED LAND MANAGEMENT (RBLM)

In recent years, the remediation of contaminated sites has received considerable attention on the national and international levels in specialized technical works and at various technical-scientific conferences to identify the most effective approaches and solutions. Through proper planning and a policy that complies with the standards and expectations of human communities under the constraints of environmental protection, the developed strategies seek to reduce the impact that contaminated sites have on humans and the environment.

When environmental damage is suspected or proven, managing contaminated sites is a tool to prevent and mitigate any negative effects as well as to reduce/avoid any threats that may exist (to human health, water, soil, habitats, food, biodiversity, *etc.*). The term "contaminated site management" refers to the integration of all conceptual components and guiding principles stated in a management strategy into a wide range of research activities, investigations, knowledge, design, and management of state authority activities, of holders of contaminated sites, from identification to rendering in non-restrictive use.

An in-depth examination of environmental issues is necessary for the management of contaminated sites, which is based on [25]:

o Human and technical resources for the management of contaminated sites in the specialized administrative structures of the Ministry of Environment and environmental agencies.

o Specific legislative framework (strategy, laws, guidelines, procedures and norms).

o Principles and objectives.

o National inventory of contaminated sites.

o Management plan dedicated to each distinct situation.

o Financing instruments.

o Informing, participating, and consulting the public in decision-making.

It is now possible to support the Management of Contaminated Sites' objective of sustainable solutions through the concept of Risk-based Land Management. The quantification of the risk associated with site pollutants and the assessment of specific risks to human and environmental health are both considered in risk-based land management. Therefore, the concept of Risk-Based Land Management (Management of Contaminated Land Based on Risk Assessment) represents a mechanism that aims to facilitate the integration of two important choices in the multi-criteria decision-making process of rehabilitation/decontamination of contaminated land: those of the time required for rehabilitation/decontamination and that of the type of solution chosen to achieve the proposed goal (Fig. **4.4**).

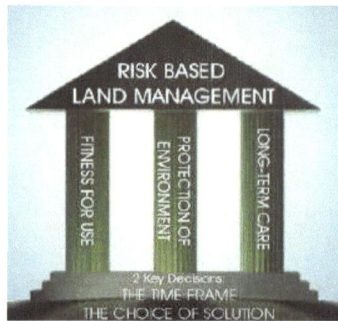

Fig. (4.4). The concept of Risk Based Land Management [26].

Three key elements that make up the RBLM concept's foundation should be considered during the decision-making process:

1. **Suitable for use** - depending on the kind and degree of contamination as well as the activity performed on the land, contaminated land can be used for a variety of purposes.

2. **Environmental protection** - the preservation and improvement (if possible) of the natural environment's quality as well as the prevention or reduction of adverse effects on the ecosystem's and biodiversity's health.

3. **Long-term care** - long-term care is required if soil contamination persists even after using a remedial strategy. The applied solution must be monitored and controlled to make sure it remains effective and that any limitations on future land use decisions are taken into consideration [26].

Environment Site Assessment and Planning

Before planning how to use, reuse, restore, or remediate the land, the environmental assessment of a site aims to ensure that the current condition of the specific site is identified, as well as any potential environmental effects that the development of a project or program related to the targeted site may have.

A site may have one or more "source" areas for pollution, such as sludge spreading areas, storage areas, discharge areas, incineration areas, industrial landfills, temporary waste storage facilities, landfills, above-ground storage tanks, underground storage tanks, *etc.*

It is necessary to carry out certain procedures to identify the type or types of existing contaminants, the environment in which they are present (soil, underground water, or both), the affected surface, or the volume that needs to be treated, before making any decision to move forward with the decontamination of a potentially polluted site. A series of analyses and evaluations of the geological environment must be done before applying a treatment in its actual form.

The process of determining the presence of pollutants in the geological environment, their spatial delimitation, determining their concentration, and determining their relationship with the mineral matrix and the geological environment's structure is referred to as **Investigation**. The methods employed, which are unique to any geological environment investigations, are tailored to the problem being investigated as well as the level of geological environment knowledge that is desired. In addition, the investigation will identify all pollutants that are currently present in the study area, establishing their spatial distribution both horizontally and vertically up to the point of identifying unaffected natural conditions, and evidencing how they got into the geological environment. It is crucial to determine how the pollutant interacts with the mineral matrix of the rocks, with underground water, and with elements of tectonics and stratigraphy.

Any technique used to measure, compute, model, forecast, or estimate the presence of a pollutant in the geological environment is referred to as an **Assessment**. The evaluation techniques include filtering primary data, statistical analysis, calculation programs to highlight elements, modeling of physical or geochemical fields, modeling of hydrogeological processes, modeling of pollutant transport, *etc.* There are knowledge limitations with the methods and techniques for investigating the geological environment. The density of observation points in relation to the map's scale serves as a categorization that restricts our understanding of the geological environment.

Finding the changes and harm that pollution has done to the soil and subsoil is the main goal of the investigation and assessment of the geological environment. The research and characterization of the contaminated area from an ecological, technical, and economic point of view using pedological, geological, hydrogeological, geochemical, and geophysical investigation works are the specific objectives of both the investigation and assessment of the geological environment [27].

The identified concentration levels of the contaminants in soil are compared to the natural background in the surrounding areas and to thresholds established at the national level, depending on the land use, to determine the intensity of pollution in a contaminated site (agricultural, residential, commercial, industrial, and others).

The following steps should be done during the investigation and evaluation of the geological environment's pollution:

o Analysis and interpretation of existing data.

o Investigation and preliminary assessment.

o Detailed investigation and assessment.

o Drawing up a geological report for investigation and evaluation of the pollution of the geological environment.

The following factors should be considered when organizing and carrying out the task of collecting representative samples of polluted soil:

o Ensuring that these samples are representative in order to accurately characterize the conditions in the study area.

o Transportation and preservation of soil samples in order to get them ready for the various physico-chemical tests that will determine the parameters being studied.

Fig. (**4.5**) illustrates a flowchart of the required activities.

A reconnaissance in the field and a sampling plan created using the information gathered there, including information on the site's past or present use, are the two stages of the preliminary investigation. Typically, no samples are taken during the field reconnaissance phase, but a summary evaluation can be done to understand the type of soil, the impact of each pollution source, and the identification of the necessary equipment for the activities that will follow. The sampling locations will be chosen so that each pollution source's influence can be evaluated, taking into account the types of sources and pollutants, the uniformity of the relief, and the characteristics of the predominant pollutants.

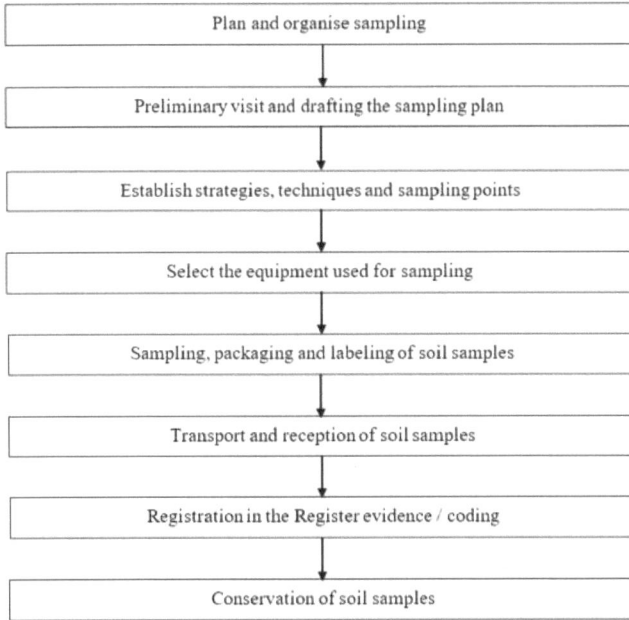

Fig. (4.5). Flow diagram of the soil sampling process.

The sampling depth will be determined in accordance with the sampling objectives and is subject to modification while field sampling is being conducted. Typically, two different depths are used for soil sample collection (collection depths of 0-20 cm for surface samples and 20-40 cm for deep samples). The soil surface should be analyzed independently of the depth layers whenever it is hypothesized that a source of pollution could represent a significant source or when the pollution's depth distribution is estimated.

The number of samples needed will be determined by the size of the potentially polluted surface being investigated and by the types of pollution present.

The Conceptual Site Model Development is carried out after preliminary planning to identify the key characteristics of the studied area, and it results in a comprehensive report that contains all the information required for a decision to remediate or reconstruct the affected geological environment.

CONCEPTUAL SITE MODEL DEVELOPMENT (CSM)

In the framework of Human Health Risk Assessment, for a better understanding of the exposure scenario, the development of a Conceptual Site Model (CSM) is recommended in the case of contaminated sites during the decision-making process, after the planning and execution of environmental investigations (see section Environment Site Assessment and - Planning).

Since CSM identifies the sources of hazards present at the site, the receptors (people, the environment, and infrastructure) who may be impacted, as well as the exposure pathways, it is useful for the assessment of contamination risks at any site. CSM contains data that is both general and site-specific. Site characteristics include a wide range of data relating to geology, hydrogeology, meteorology, actual or possible receptors, and contaminant source characteristics. This involves knowledge related to site operations and previous investigations [28, 29].

Plans for sampling, analysis, and monitoring, as well as risk assessment, are developed on CSM:

o Identifies the main sources of environmental discharges or potential environmental discharges from an industrial or commercial activity (*e.g.* disposal of wastes to soil or water, airborne releases from an industrial operation, or acid drainage from a mine site).

o Demonstrates the potential mobility of discharges or contaminants at the point of release in the environment (*e.g.* chemicals from an above ground storage area may move down into soil or water; petrol from an underground tank may migrate off-site and affect near neighbours; or contaminants from fill material may move into groundwater).

o Defines the various receptors that might be exposed to contaminated media (*e.g.,* birds, mammals, fish, plants, humans), and;

o Provides a list of the possible interactions between various receptors and contaminants (*e.g.,* potential exposure pathways through ingestion of contaminated surface or groundwater, ingestion of contaminants in soil or food, and direct contact with contaminated soil or water) [30].

The CSM is a flexible tool that can be expanded or contracted as needed during development based on the findings of environmental investigations. Thus, only the existence of all four of the previously mentioned components makes an exposure pathway complete.

Here are many ways to represent a CSM, including narrative, diagrammatic, table format, picture, flow charts, exposure pathway model, fate and transport model, and interactive electronic or virtual 3D (Fig. **4.6**) [30].

EXAMPLE FOR PATHWAY-EXPOSURE CSM

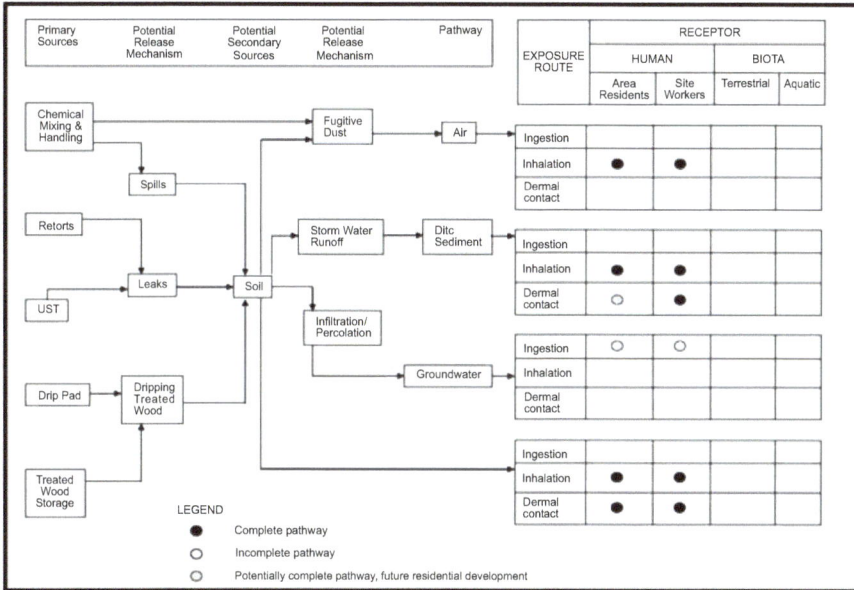

Fig. (4.6). Example of CSM – flowchart [31].

The Conceptual Site Model (visual representation) of the petroleum migration pathways from source to receptors is shown in Fig. (4.7).

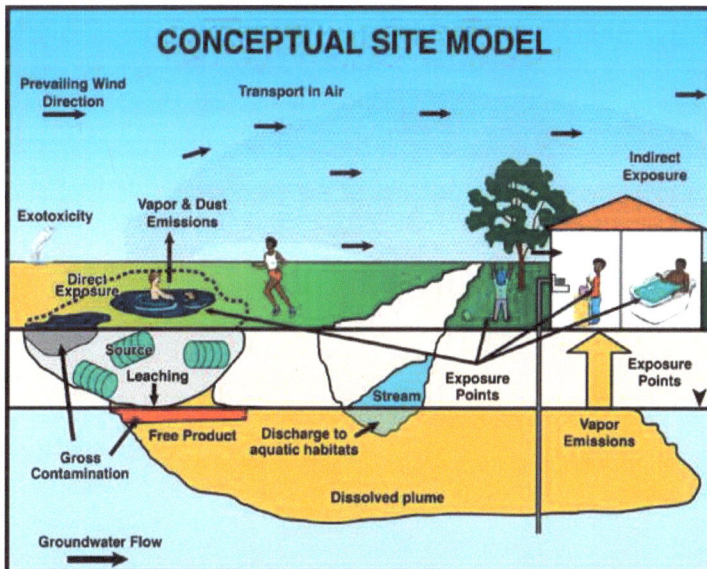

Fig. (4.7). CSM evidencing the migration pathways of petroleum from source to receptors [32].

People can directly come into contact with heavy metals (toxic polluting substances), as previously discussed (see paragraphs Soil Pollution and Health and Conceptual Site Model Development (CSM)), through diet, livestock, sea, and drinking water, by inhaling polluted air, or by occupational exposure at work [33]. This is almost always the cyclical order in which heavy metal contamination occurs: industry, atmosphere, soil, water, food, and then humans [34]. There are different pathways for heavy metals to enter the human body. Some heavy metals, like lead, cadmium, manganese, and arsenic, can enter the body through the digestive tract, or through the mouth when the receptor is eating food, drinking water, or consuming other beverages. While some, like lead, can be absorbed through the skin, or others, like smoke, can enter the body through inhalation.

Human Health Risk Assessment is the process of estimating the nature and likelihood of adverse health effects in people who may be exposed to chemicals in contaminated environmental media, either now or in the future.

Four main steps are included in Human Health Risk Assessment:

o Step 1 - **Hazard Identification**

Risk evaluator(s) consider whether and under what conditions a stressor has the potential to harm people and/or ecological systems.

o Step 2 - **Dose-Response Assessment**

To ascertain the quantitative relationship between exposure and effects, risk evaluators analyse the achieved data and information.

o Step 3 - **Exposure Assessment**

Risk evaluator(s) look at what is known about the frequency, timing, and levels of contact with the stressor after steps 1 and 2 have been completed.

o Step 4 - **Risk Characterization**

Risk estimation and risk description are the two main parts of risk characterization.

The term "**Risk Estimation**" refers to the process of comparing information about the potential receptor's expected effects with the estimated or measured exposure level to each stressor [35].

The information in the "**Risk Description**" is crucial for interpreting the risk results. Among them are:

o If adverse effects on the threatened plants and animals are anticipated

o Comparisons of relevant qualitative data

o How the assessment might be impacted by uncertainties (data gaps and natural variation) [35].

Exposure assessment is the process by which: **(1)** potentially exposed populations are identified; **(2)** potential pathways of exposure and exposure conditions are identified; and **(3)** chemical intakes/potential doses are quantified. Exposure may occur by ingestion, inhalation, or dermal absorption routes.

The process of identifying potentially exposed groups of people **(1)**, identifying potential exposure pathways **(2)**, identifying exposure **circumstances**, and quanti-fying chemical intakes and potential doses **(3)** is known as Exposure Assessment. Ingestion, inhalation, and dermal absorption are all possible **routes** of exposure.

The term "dose" describes the quantity of a chemical that crosses an individual's external boundary. The rate of intake (such as ingestion or inhalation) or uptake *(i.e.,* dose) depends on the concentration of the contaminant. The amount of a chemical that could be ingested, inhaled, or deposited on the skin is known as the potential dose. The absorbed dose is the quantity of chemical that enters the body through the skin, lungs, or digestive system. The potential dose from animal feeding studies or the absorbed dose from pharmacokinetic studies, followed by intraperitoneal or other injected delivery into the test animal, serve as the toxicological foundation for risk assessment, respectively. One method for calculating potential dose (PD) is as follows:

$$PD = C \times IR \qquad (4.1)$$

where:

PD = Potential dose (mg/day);

C = Contaminant concentration in the media of interest (mg/cm2, mg/m3, mg/g, mg/L); and

IR = Intake or contact rate with that media (cm2/day, m3/day, g/day, L/day) [36]

Risk Caracterisation

• Calculating Exposure Doses

To determine the exposure dose caused by coming into contact with a contaminated medium, the general equation shown below is used [37]:

- Generic Exposure Dose Equation

$$Dose = C \times IR \times AF \times EF / BW \tag{4.2}$$

Where,

D = Exposure dose

C = Contaminant concentration

IR = Intake rate of contaminant medium

AF = Bioavailability factor^{-1}

EF = Exposure factor

BW = Body weight

The general formula for estimating exposure dose is followed by examples for various exposure pathways in the following paragraphs:

- *Exposure through accidental ingestion of contaminated soil*

$$Dose_{s.i.} = (C_{cont} \times IR \times FI) \times (BW^{-1}) \times ((ED \times EF) \times AT^{-1})) \times CF \tag{4.3}$$

Where:

Dose s.i. – mg/kg/day – Exposure dose

C_{cont} – mg/kg$_{d.w.}$ - The concentration of the contaminant in the soil

IR – mg soil/day - The ingestion rate

FI – unitless - The fraction ingested from the contaminated source

- *Exposure through dermal contact*

$$Dose_{d.c.} = ((C_{cont} \times SA \times AF \times ABS) \times BW^{-1}) \times ((EF \times ED) \times AT^{-1})) \times CF \tag{4.4}$$

Where:

Dose d.c. – mg/kg/day – Exposure dose

C_{cont} – $mg/kg_{d.w.}$- Concentration of the contaminant in the soil.

SA - cm^2/day - The surface area of skin available to contact soil-to-skin.

AF - mg/cm^2 - Adherence factor.

ABS- unitless - Traction absorbed across the skin.

Where:

BW represents the body weight,

CF – Conversion factor = 10^{-6} kg/mg,

EF – Exposure frequency (days/year),

ED - Exposure duration (years),

AT the average time (days).

REMEDIATION TECHNOLOGIES

The following categories represent a classification of soil and groundwater treatment technologies (Fig. **4.8**) [38]:

o Physical treatment technologies

o Chemical treatment technologies

o Thermal treatment technologies

o Biological treatment technologies

Any of these can be divided into *In-Situ* and *Ex-Situ* technologies. On-Site and Off-Site technologies are subcategories of *Ex-Situ* technologies. More details in this regard will be illustrated in the next paragraphs.

The **physical treatment technologies** involve moving and concentrating the pollutants to the extraction points while utilizing the ground's already-existing drivers (air, gas, and moisture). These methods combine off-site, and in-site and on-site actions (those pertaining to the treatment of water and air used to transport pollutants). The off-site actions involve treating highly contaminated products, which must be done in the proper locations for the process's equipment and security.

Biological remediation	Chemical remediation	Physical remediation	Thermal remediation
Soil bioremediation	Electrochemical remediation	Soil covering/soil encapsulation	Soil incineration
Soil bioventing	Soil flushing and soil washing	Soil excavation	Soil pyrolysis
Soil phitoremediation	Chemical oxidation	Soil mixing	Soil vitrification

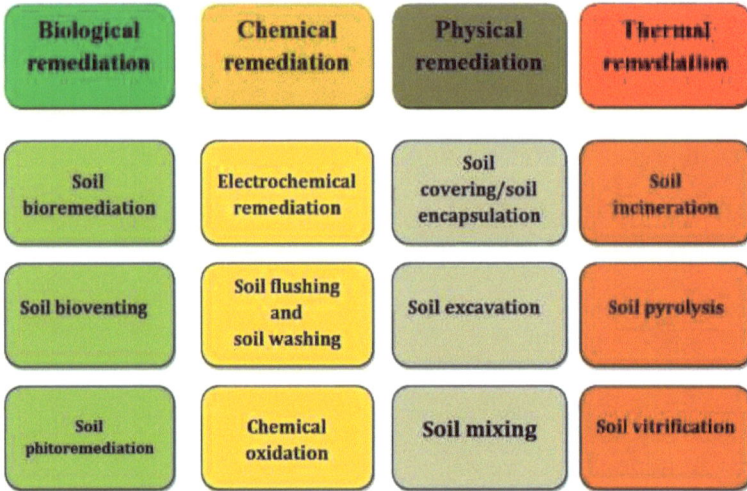

Fig. (4.8). The main technologies for the remediation of contaminated sites.

Toxic compounds can be eliminated, fixed, or given more mobility and neutralization using **chemical treatment technologies**. These technologies are based on some chemical reagents' capacity to fulfill their intended purpose. Chemical treatment techniques have evolved in two directions that, while acting through apparently distinct processes, are comparable in terms of the installations and equipment used:

o Techniques that ensure the immobilization and transformation of pollutants.

o Techniques for mobilizing and removing (washing) contaminants from the underground environment.

High temperatures are used in **thermal treatment technologies**, specifically to destroy organic pollutants (*e.g.,* hydrocarbons). Thermal desorption, which increases pollutant mobility, volatilization, and the subsequent capture of residual gases for treatment, is another thermal treatment for contaminated soils. High temperatures are used for the thermal treatments, which are typically applied to soil that has been excavated (off-site).

The use of microorganisms with the capacity to convert organic pollutants primarily into CO_2, water, and biomass or to immobilize pollutants by binding in the humic fraction of the soil constitutes **biological treatment technologies**. Due to the complexity of the situations, controlling biodegradation in the field is challenging (non-homogeneous soil, multiple pollutants, alternation of seasons, *etc.*). However, it is frequently employed, particularly when used in conjunction with physical treatment technology.

Degradation typically takes place in aerobic or, less frequently, anaerobic conditions. The conditions for the growth of microorganisms must be optimized to increase the process' efficiency (oxygen supply, pH, water content, *etc.*). Soil homogenization, active aeration, humidification or drying, heating, addition of nutrients or substrates, and inoculation with microorganisms can all be used to stimulate biological activity. Compared to thermal or physico-chemical processes, biological ones demand much less energy (low costs), but they also demand longer treatment times [39].

The goal of all these technological approaches is to return the sites to a condition that is as close to the initial one as possible (for example, physical-chemical treatment). These technologies should be built on methods that don't harm the environment. The following requirements for the remediation methods must be met:

o To eliminate or neutralize contaminants.

o To have the least possible impact on the resources used (water, air, energy, capital).

o To avoid producing contaminated secondary emissions (air or process water).

o To maximize the amount of recycled material from the treated material.

o To reduce the amount of residual materials and their potential for danger.

o To avoid creating new toxic substances by removing or transferring those that already exist.

Due to the extremely wide variety of pollutants and the various areas of spread, a single technology typically is not sufficient to remediate a site. The so-called treatment strategy usually involves the application of several technologies. For instance, the removal of vapors from the soil can be combined with the pumping of underground water and the injection of air into the soil to gradually remove pollutants from both the water and the soil (physico-chemical). A single treatment unit can handle emissions from the air injector, the soil and air vapor extraction system, and other sources (chemical treatment). Another benefit of this system is the air circulation through the soil, which promotes or intensifies natural biological activity while also causing some pollutants to degrade. In some cases, air is injected into both saturated and unsaturated zones to facilitate the movement of pollutants and to promote biological (physico-biological) activity.

In general, a particular group of contaminants can be remedied using treatment technologies. However, due to certain factors like cost, public acceptance, or implementation, it might not be applicable on a large scale.

The following are some of the most crucial factors that influence the selection of the most suitable decontamination technology, when a remedial action plan is established:

o Soil characteristics

o Pollutant's type

o Land use

o Seasonal weather conditions

o Costs

o Remediation time

o The need to monitor how the decontaminated site changes over time and the need to act with specific treatments if certain undesirable phenomena start to appear.

o Identification of the action of microorganisms towards various categories of pollutants.

o The existence of the specific equipment.

o Remediation targets, desired or imposed.

o The side effects produced during the application of remediation technologies and after their application.

This final factor, which must be considered when selecting and using a contaminated soil depollution technique, is a component of the current global trend in the industry, which is represented by the idea of the Green and Sustainable Remediation approach - GSR [40]. GSR is described as follows by the Environmental Protection Agency (EPA): *"Any technique, strategy, or management plan that considers environmental, economic and social aspects, which can reduce environmental footprints, negative social economic impacts throughout the remediation process, from site investigation, remedy design, operation and management to site closure, while still meeting the regulatory requirements"* [41].

The practical situations that may exist around contaminated soils are very diverse, and this has led to the development of a wide range of techniques to treat them, some fully developed and others in the research phase. Therefore, the choice of one or the other, or a combination of them, will depend on the characteristics of the soil and the pollutant to be treated, the level of decontamination required, the economic viability and the time available to carry out the project.

As previously mentioned, there are also different classifications, one of them depending on the place of application of the technique. They are called *in situ*, when the management technique is carried out on the same site where the contaminated soil is located; it will be *ex situ*, when the deteriorated soil is excavated. If the contaminated soil is moved to other facilities for treatment, it will be *off-site*, while if there is no transport of the material, it will be *on-site*. In general, *on-site* techniques tend to be slow as the contact between the contaminant and the chemical agents is allowed by the environment and the ecosystem, limiting the possibilities of optimization. As for *ex situ* techniques, they are usually more costly due to excavation, but also more efficient and faster since contact is more efficient. Another classification is based on the objective to be followed in the process [42], which allows us to distinguish three large groups [43]: **containment techniques** that consist of physically isolating the pollutant from the environment, without acting on it; **confinement techniques** that try to prevent the migration of the pollutant in the soil by reducing its mobility by acting on its physical-chemical properties; and **decontamination techniques** whose objective is to drastically reduce the amount of pollutant in the soil by acting directly on it by means of chemical, physical and/or biological processes.

CONTAINMENT TECHNIQUES

The general purpose is to prevent or significantly reduce the migration of pollutants into the surrounding environment, including groundwater. They are based on the construction of physical barriers and are therefore quick to implement and relatively low cost as they do not require excavation of significant volumes of soil. They are often used as a first approximation in the design of the total decontamination process. They are distinguished:

• **Vertical barriers:** This aims to minimize the lateral migration of contaminants, including leached or dissolved contaminants in groundwater, by digging trenches up to 100 m deep (Fig. **4.9A**) in preferably coarse-textured and not very compacted soils that are backfilled with insulating material such as cement-bentonite mixtures or concrete [44].

• **Horizontal barriers:** Focuses on preventing or minimizing the vertical migration of contaminants present in the soil by digging horizontal trenches or boreholes that are filled with sealing material [45]. A variant of this technique is the 'dry soil barrier', whereby dry air flows through the vertical or horizontal boreholes, reaching the contaminated area, vaporizes the water in the soil, carrying the contaminant with it and exiting through the extraction wells (Fig. **4.9 B**). Once at the surface, this air is treated and dried for re-injection into the system [46].

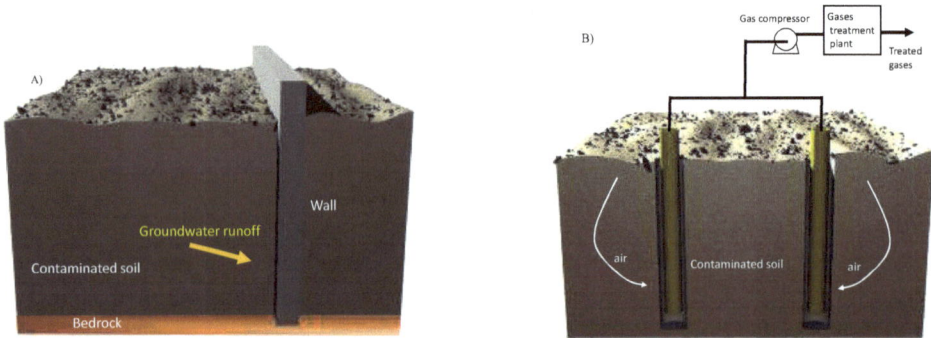

Fig. (4.9). (**A**) Schematic diagram of the vertical barrier technique, (**B**) Schematic diagram of the application of the dry floor barrier.

• **Surface sealing:** This is applied *in situ* to impoverished or superficially degraded soils to prevent the migration of light pollutants to the atmosphere and the infiltration of rainwater or runoff, as this would lead to leachate that could easily advance to deeper areas, including groundwater. For this purpose, vertical boreholes are drilled (without reaching the water table) through which sealing material such as bentonite is introduced to reduce the permeability of the medium.

• **Deep sealing:** The objective is to modify the soil structure *in situ* to decrease its permeability by controlling the migration of the contaminant into deeper layers. The plasticizing material (cement with bentonite, bentonite with organic resins or sodium silicate) is injected through separate vertical boreholes reaching different depths: up to 20 m for permanent seals and up to 30 m for temporary seals.

• **Hydraulic barriers:** The intention is to protect the groundwater, and for this purpose, it is extracted by means of wells, drains or drainage ditches dug in the vicinity of the contaminated area, causing the water table to be lowered. When the water has already been in contact with the contaminant, it is treated in the outside area and can be re-injected once it has been cleaned to minimize the negative effects associated with the extraction of the water in the environment [43].

CONFINEMENT TECHNIQUES

The general purpose is to reduce the mobility of pollutants by minimizing the chances of dispersion in the medium by employing different physical and/or chemical processes by making the substances progress to less soluble and toxic forms (stabilization), to be integrated into a glassy matrix (vitrification) or to be impermeable to water (injection of solidifying agents). The limitations of these techniques are mainly found in the treatment of organic substances and pesticides, as they are reduced to solidification with asphalts and vitrification [47].

• **Physic-chemical stabilization:** This consists of reacting the pollutants present in the soil with chemical reagents, reducing their solubility in the soil and limiting the possibilities of leachate formation. It is often carried out *ex situ*, as it improves the pollutant-chemical reagent contact, requiring the prior elimination of the coarse fraction. The soil to be treated is placed in mixing tanks with the stabilizing agents (cement and phosphates or alkalis) dissolved in water that will react with the pollutant. For example, a pH increase can be caused that will favor the precipitation or immobilization of certain heavy metals [47, 48]. Subsequently, they are separated from the soil and the soil is subjected to regeneration to be able to return it to its area of origin. If it is not viable, it is deposited in controlled landfills.

• **Injection of solidifying agents:** This technique is a combination of physic-chemical stabilization and deep sealing, as it is based on injecting chemical agents into the soil (*in situ*) through wells drilled into the soil, physically encapsulating the contaminants in a stable, water-impermeable matrix that limits their mobility in the medium [49]. The most used chemical reagents are inorganic compounds such as cement or organic compounds such as bituminous substances, polyethylene or paraffin [50] and it is a viable technique for treating soils contaminated with inorganic substances [50].

• **Vitrification:** It is a destructive procedure that performs the thermal stabilization of the system by forming a glassy matrix that will include the pollutants present, generally inorganic, and mainly heavy metals (Hg, Pb, Cd), although it is also viable for others (As, Ba, Cr and cyanides). The application of high temperature also makes it suitable for the management of organic pollutants, as they will be pyrolyzed and/or oxidized. To be effective, the soil must be rich in silica (base of the matrix) and in alkaline oxides (Li, Na, K,...) that will act as stabilizing agents. It is also interesting that the humidity of the soil is low in order not to lose energy by evaporation and that the clay and silt content is low as it would retain the water evaporated during the heating process. Depending on the site, vitrification can be carried out *ex situ* in an oven at temperatures between

1100°C and 1400°C [51] or *in situ* using graphite electrodes to reach the required 1600°C to 2000°C through electric current [52].

DECONTAMINATION TECHNIQUES

They aim to eliminate, and when it is not possible to minimize, the concentration of contaminants present in soil. They comprise a wide range of processes that can be grouped into three main groups: **Physical-Chemical**, which relies on the physical and chemical properties of the medium to remove the pollutant from the affected soil; **Biological**, which is based on the degradative capacity of microorganisms and plants to remediate the soil; and **Thermal**, which makes use of high temperatures to reduce the amount of pollutants.

Physico-Chemical Techniques

• **Extraction:** This is applied *in situ* and allows the pollutants to be separated by integrating a purification process as subsequent treatment. They are generally simple and are applied in permeable soils, and when the pollutant is mobile, *i.e.*, it has not been adsorbed by the medium. The wide variety of fluids that can be used and the particularities of the system employed make it possible to distinguish between multiple variants, such as **Air Stripping** (Fig. **4.10**), which uses vertical or horizontal extraction wells to generate a pressure gradient and extract volatile or semi-volatile pollutants in the gas phase that are not adsorbed by unsaturated soils [53 - 55]; **Water Extraction** appropriate when the contaminant is present in aquifers or in the unsaturated zone [49] as the water is pumped through wells and treated at the surface; **Free Phase Extraction** used to remediate soils contaminated with hydrocarbons that are immiscible with water due to the distinct layer above the water table it forms; and **Dense Phase Extraction** used when the contaminant is denser than water, poorly soluble, difficult to degrade naturally and accumulates below the water table.

• **Washing:** This is generally applied *ex situ* and consists of bringing the soil to be remediated, previously screened, into contact with chemical agents to extract and solubilize the contaminants present, hence it can also be called solvent and acid extraction. The solvent carries away the contaminants and is then separated from the soil by evaporation. Both the vapor and the soil must be regenerated, the former by contact with other solvents or distillation and the latter by continuous washing processes to remove chemical residues. The wide variety of possible washing agents facilitates the control of many soil properties such as pH, texture, cation exchange capacity, mineralogy or organic matter content [55, 56], making the technique feasible both for petroleum hydrocarbons and semi-volatile organic compounds when using surfactants [57] as well as for inorganic substances such as heavy metals and cyanides when using chelating substances such as ethylenediaminetetraacetic acid (EDTA), citric acid, nitrilotriacetic acid (NTA) and hydrochloric acid [58, 59].

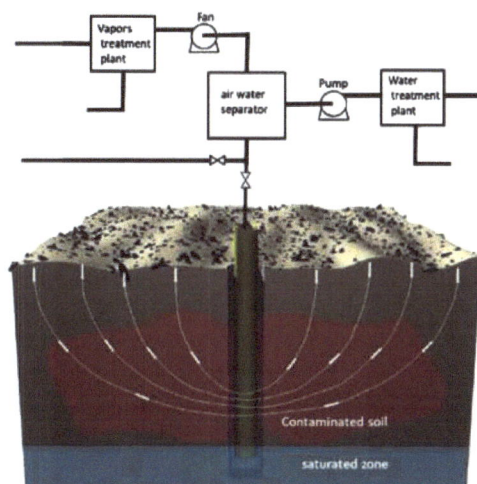

Fig. (4.10). Illustrative diagram of the rehabilitation of a floor using the air extraction technique.

• **Flushing:** This consists of flooding a specific area of contaminated soil with an aqueous solution (they can also be injected) capable of dragging and transporting the contaminants present in the soil. The mixture is then collected by pumping through extraction wells. Once at the surface, the mixture is treated and can sometimes be reused [47, 60]. This technique is very efficient for remediating permeable soils with a high bulk fraction contaminated by inorganic substances, including radioactive elements. However, the transport of the plume within the soil must be carefully controlled to avoid dispersion of the contaminant to unaffected areas.

• **Electrochemically Assisted Soil Remediation (EKR):** An *in-situ* technology that applies an electric field to cause the migration and removal of organic and inorganic contaminants in soil and sediments, including fine-grained and low hydraulic conductivity soils. For this purpose, a direct current, typically mA/cm^2 (cross-sectional area), is conducted from a negatively charged electrode or cathode to a positively charged electrode or anode [61, 62].

In practice, EKR is carried out in two ways: either by directly burying the two electrodes in the ground and bringing them into contact with each other, or indirectly by drilling boreholes in which the electrodes are placed next to an electrolyte solution (Figs. **4.11A** and **B**) respectively). Both cases promote the migration and enrichment of contaminants near the electrode area, but decontamination will only occur if chemicals such as surfactants and weak acids

[63] or biological components [64] are introduced into the soil or electrochemical solution. In addition to removing and extracting contaminants, the injected electrolyte solution can be used to improve the electrical conductivity of the soil, control the polarization of the electrodes, and extend the lifetime and efficiency of the system.

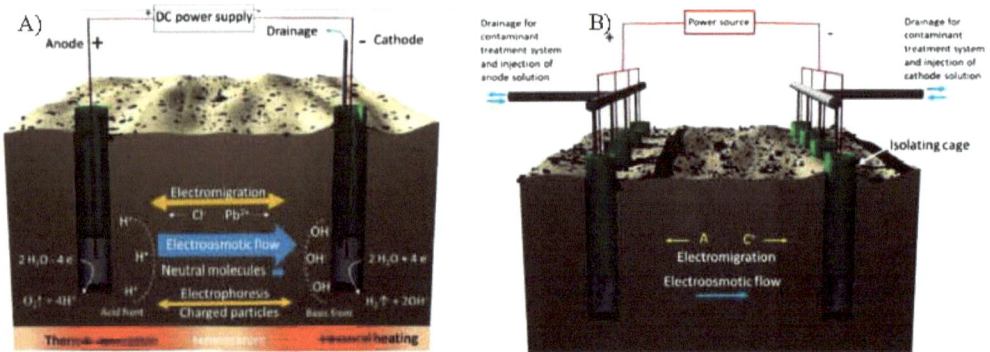

Fig. (4.11). Processes in electrochemical soil remediation: **A)** Direct contact, **B)** Indirect contact.

• **Addition of amendments:** This involves adding chemical agents in a solid state so that once mixed with the soil, they can transform the contaminants (contact can be enhanced by irrigation). It is a procedure used to remediate soils contaminated by salts or heavy metals, *i.e.*, desalination of sodium-rich soils remediated by cation exchange of calcium compounds [65, 66]. In the case of the presence of toxic metals such as P, U, As, Zn, Ni Sr, Cu and Cd, the addition of zeolites, phosphates, iron minerals, bentonites, calcium hydroxide, compost or yeasts that help to immobilize pollutants is often used [66, 67].

• **Active permeable barriers:** This is applied *in situ* and consists of constructing a semi-permeable wall or screen under the ground, through which contaminated groundwater circulates naturally or forcibly while contaminants are trapped, or their dangerousness is reduced [68 - 70]. The mechanisms of immobilization and transformation of pollutants in active barriers can be classified into three [68]: chemical and/or biological reactions leading to the decomposition and degradation of the pollutant into non-toxic or harmless substances; physical or physico-chemical adsorption processes by complex formation; and chemical precipitation leading to insoluble compounds. Examples of successfully applied barriers are those using zero-valent metal elements such as iron that led to redox processes of chlorinated solvents such as trichloroethane or tetrachloroethane [71], of trace and radioactive metals [72] and of inorganic pollutants such as nitrates and sulphates.

• **Fracturing:** involves forming cracks in soil zones with low permeability by injecting pressurized air to open up the possibility of remediating the soil with other techniques. Fracturing can be a pneumatic, enhanced blast, hydraulic and the Lasagna process™ [73, 74].

• **Compressed air injection:** The aim is to cause the volatilization of contaminants in the soil or groundwater by injecting air using pre-dug wells. The vapor is then displaced to the unsaturated zone, where it is collected by extraction wells and purified [75]. To increase the efficiency of the process, it is advisable to eliminate the existing free phase before proceeding with air injection and to restrict the technique to sufficiently permeable soils, with low humidity and for volatile and semi-volatile substances of low molecular weight such as xylene, benzene, toluene, carbon tetrachloride, trichloroethane, methyl chloride, among others [76].

• **Ultraviolet oxidation:** It can be applied both *in situ* and *ex situ* and consists of performing oxidation/reduction reactions on pollutants to obtain non-hazardous or less toxic compounds that are more stable, less mobile, or even inert [77]. The oxidizing compounds are introduced through wells to facilitate contact with the contaminated area and once the reaction is complete, the mixture is pumped out to be treated [78]. The most used oxidants are potassium permanganate, persulphate, hydrogen peroxide, and ozone [59, 79, 80].

• Recent studies have shown that the use of biochar as an amendment of polluted soil can lead to the reduction of pollutant concentrations, such as heavy metals, through different intermolecular interaction mechanisms: physical adsorption, redox reactions, electrostatic interaction, complexation, ion exchange and co-precipitation [81]. The ability of biochar to reduce pollutant concentrations depends on several factors that are closely related to the type of material from which the biochar is obtained and the technical parameters (temperature level, pressure, retention time, *etc.*) at which it is produced. At the same time, within the use of this method, mechanisms to reduce the aging of biochar must also be taken into account because part of the initially immobilized pollutants can become available again after the degradation of the biochar [81].

Biological Treatment

Also known as bioremediation, they use natural biological activity to degrade organic pollutants or reduce the toxicity of inorganic compounds, including toxic metals, through reactions that are part of their metabolic processes [82]. These processes are carried out by bacteria, fungi, and plants, and to be efficient, a comfortable situation must be achieved by controlling parameters such as the type and concentration of pollutants present, the presence of sufficient oxygen in aerobic processes, as well as nutrients, humidity, temperature, pH or bioaugmentation. They are simple, economical, and environmentally friendly [82] and technologies such as bio-correction/bioremediation, phytoremediation, bioventing, bio-piles, composting, landfarming or biological treatment in the suspension phase are differentiated [83]. It should be noted that they can be carried out *in situ* or *ex situ*, although *in situ* technologies generally require more time for their development, are subject to the heterogeneity of soil and aquifer characteristics and their effectiveness is more difficult to verify.

On-Site Treatments

• **Bio-correction/bioremediation:** It is applied both for soil remediation and groundwater decontamination and is because indigenous or inoculated microorganisms use the pollutants as a source of energy and food to continue their life process and give rise to harmless substances. The growth of micro-organisms requires the presence of electron donors and acceptors, which are necessary to oxidize and reduce organic compounds (pollutants), in this way, they will be chemically transformed and become the source of carbon used by the cellular constituents as well as generating the necessary energy that makes the synthesis and maintenance of biomass possible. If they are no longer able to grow from them, they can continue to transform them if an alternative growth substrate is provided. In addition, micro-organisms need a source of nutrients (N, P, K, S, Mg, Ca, Mn, Fe, Zn, Cu and trace elements) and suitable environmental conditions in terms of pH, temperature, electrical conductivity, water content, oxygen content and redox potential [83, 84].

• **Phytoremediation:** This is based on the capacity of certain plant species to survive in environments contaminated with heavy metals and organic substances while cleaning it through different mechanisms such as those based on contaminant containment (phyto-stabilisation or phyto-immobilisation) or elimination (phyto-extraction, phyto-degradation, phyto-volatilisation and rhizo-philisation). In addition, plants help to prevent or minimize soil erosion by wind and water [85]. The success of this technique will depend on an appropriate selection of the plant species to be planted and the need for amendments (organic matter, chelating agents, lime, *etc.*) to improve soil properties and promote plant survival and growth.

• **Bioventing:** This is carried out *in situ* and consists of the injection through wells of air (or oxygen) and, if necessary, nutrients, in order to stimulate the microbial activity that will degrade the pollutants concentrated in the unsaturated zone [86], Any aerobically biodegradable substance is susceptible to treatment by this process, such as petroleum hydrocarbons [87], especially those of medium molecular weight, as light ones will tend to evaporate over time and are best treated by vapour extraction, while heavy ones would take a long time to biodegrade.

Off-Site Treatments

• **Landfarming: Landfarming** involves the disposal of contaminated soil, previously excavated, in the form of a thin layer, no more than 1.5 m high, on controlled soil. To stimulate aerobic microbial activity, aeration stages are carried out. This isolated and controlled arrangement allows for the addition of allochthonous degrading bacteria to accelerate the process, nutrients, minerals, and moisture [88], as well as the provision of control systems for temperature, pH and other parameters. It has proven to be efficient for the treatment of soils with hydrocarbons and explosives such as TNT.

• **Bio-piles:** Excavated soil is arranged in successive piles on impermeable soil to prevent seepage, where aerobic microbial activity is stimulated by aeration and nutrient addition. It is an efficient technique for petroleum compounds, halogenated and non-halogenated volatile organic compounds, semi-volatile organic compounds, and pesticides [89, 90]. In contrast to landfarming, the introduction of air is not done by ploughing but by forcing air circulation by injecting or extracting air through perforated ducts placed inside the material. The treatment is carried out for short periods of time, ranging from a few weeks to a few months, during which time the piles are usually covered with impermeable material to prevent the release of pollutants into the atmosphere and the infiltration of rainwater.

• **Composting:** This is a biological process widely used to cause aerobic and anaerobic degradation of toxic organic compounds by stimulating the biodegradation activity of indigenous microorganisms under thermophilic conditions (12°-18°C). In this system, the contaminated excavated soil is mixed with animal and vegetable waste such as manure, dung, straw, wood chips, *etc.*, which provide porosity with adequate amounts of carbon and nitrogen to carry out the decontamination process. In this system, it is common to observe an increase in temperature in the matrix of the piled material due to the metabolic activity and cooling at the end of the activity. It is an effective technique to reduce the concentration of explosives (TNT, RDX and HMX) [91, 92], polycyclic aromatic hydrocarbons [93], petroleum hydrocarbons [94], chlorophenols [95] and pesticides [96].

• **Biological sludge:** *Ex situ* treatment that brings excavated and screened contaminated soil into contact with water and other additives in a controlled bioreactor to optimize the process. The result is a sludge where solids are kept in suspension, facilitating contact between the biodegrading micro-organisms and the contaminants. As the process is carried out in a closed device, the availability of substrates, nutrients, oxygen, temperature, pH, humidity, and homogenization of the mixture can be controlled by moderate agitation. In addition to biodegradation, adsorption/desorption, dissolution, precipitation, ion exchange, complexation, oxygen transfer, volatilization and particle size reduction can occur [97].

Heat Treatment

• **Incineration:** This is an *ex-situ* and destructive decontamination procedure that uses high temperatures to destroy contaminants. Soil is subjected to temperatures of around 1000°C in combustion furnaces to oxidize and volatilize organic pollutants. It is a technique that generates waste gases and ashes, organic compounds (polycyclic aromatic and sulphurous hydrocarbons, oxygenated compounds, nitrogenous aromatic compounds, *etc.*) and inorganic compounds (volatile heavy metals, CO_2, NO_x, SO_x) [98] that must be purified.

• **Thermal desorption**: This is an *ex-situ* technique focused on extracting contaminants by volatilization while avoiding their oxidation. It can be carried out at low (90-320°C) and high (320-560°C) temperatures. In the case of low-temperature thermal desorption, the soil retains both its physical properties and the organic components present, making remediation feasible. It is often used to remediate soils contaminated with non-halogenated volatile organic compounds, fuels and, in some cases, semi-volatile organic compounds. High-temperature thermal desorption can treat the above substances in addition to polycyclic aromatic hydrocarbons, PCBs, pesticides, and volatile heavy metals such as Hg and Pb [99 - 101].

CONCLUDING REMARKS

The current chapter highlights the importance of developing a Conceptual Site Model (CSM) as an essential tool in the contamination risk assessment methodology being evidenced the sources of pollution, exposure routes and potential receptors are identified. Thus, in the decision to remediate a contaminating site, all technical-scientific and economic information are considered in order identifying the remediation method or methods applied in order to eliminate or reduce the level of contamination. The chapter presents the most common decontamination methods used in order to highlight that, for the complete decontamination of a site, especcially when it is a about a complex contamination, often it is necessary to apply several remediation methods either sequentially or simultaneously.

REFERENCES

[1]   L. Anicai, C. Bâsceanu, M. Dutu, S. Chineata, O. Anicai, D. Staniloae, and R. Dumitrache, *Integrated management of contaminated soils* Ed. PRINTECH: Bucureşti, 2010.

[2]   "European environment agency", Available at: https://www.eea.europa.eu/data-and-maps/figures/overview-of-progress-in-the-management-of-contaminated-sites-in-europe/csi015-fig01noprojec-june 07.eps

[3]   "European environment agency", Available at: https://www.eea.europa.eu/data-and-maps/figures/breakdown-of-activities-causing-local-soil-contamination/csi015-fig05-july07.eps

[4]   E.S. Rubin, *Engineering and Environment* Mc Graw Hill: NY, 2001, p. 696.

[5]   D.M. Cocârţă, *Human Health Risks from Non-Renewable Energy* Polytechnic Press: Polytechnic University of Bucharest, p. 114, 2017.

[6]   T. Münzel, O. Hahad, A. Daiber, and P.J. Landrigan, "Soil and water pollution and human health: what should cardiologists worry about?", *Cardiovasc. Res.,* vol. 19, no. 2, pp. 440-449, 2022.
[http://dx.doi.org/10.1093/cvr/cvac082]

[7]   C. Dumitrescu, D.M. Cocârţă, and A. Badea, "An integrated modeling approach for risk assessment of heavy metals in soils", In: *UPB Sci. Bull.* vol. 74. Elsevier, no. 3, pp. 217-228, 2012.

[8]   L. Järup, "Hazards of heavy metal contamination", *Br. Med. Bull.,* vol. 68, no. 1, pp. 167-182, 2003.
[http://dx.doi.org/10.1093/bmb/ldg032]

[9]   S. Martin, and W. Griswold, "Human health effects of heavy metals", *Environ. Sci. Tech. Bri. Citi.,* vol. 15, pp. 1-6, 2009.

[10]   M. Markowitz, "Lead poisoning", *Pediatr. Rev.,* vol. 21, no. 10, pp. 327-335, 2000.
[http://dx.doi.org/10.1542/pir.21.10.327]

[11]   M. Alina, A. Azrina, M.A.S. Yunus, S.M. Zakiuddin, M.I.H. Effendi, and R.M. Rizal, "Heavy metals (mercury, arsenic, cadmium, plumbum) in selected marine fish and shellfish along the straits of Malacca", *Int. Food Res. J.,* vol. 19, no. 1, pp. 135-140, 2012.

[12]   A. Bernard, "Cadmium & its adverse effects on human health", *Indian J. Med. Res.,* vol. 128, no. 4, pp. 557-564, 2008.

[13]   K. Shekhawat, S. Chatterjee, and B. Joshi, "Chromium toxicity and its health hazards", *Int. J. Adv. Res.,* vol. 7, no. 3, pp. 167-172, 2015.

[14]   R.S. Hillman, „Hematopoietic agents: Growth factors, minerals, and vitamins", In: J.G. Hardman, L.E. Limbird, A.G. Gilman, editors. Goodman & Gilman's the Pharmacological Basis of Therapeutics. 10th ed. New York: McGraw-Hill; 2001. pp. 1487-1518.

[15]   S.L. O'Neal, and W. Zheng, "Manganese toxicity upon overexposure: A decade in review", *Curr. Environ. Health Rep.,* vol. 2, no. 3, pp. 315-328, 2015.
[http://dx.doi.org/10.1007/s40572-015-0056-x]

[16]   K.J. Klos, M. Chandler, N. Kumar, J.E. Ahlskog, and K.A. Josephs, "Neuropsychological profiles of manganese neurotoxicity", *Eur. J. Neurol.,* vol. 13, no. 10, pp. 1139-1141, 2006.
[http://dx.doi.org/10.1111/j.1468-1331.2006.01407.x]

[17]   E.C. Brevik, L. Slaughter, B.R. Singh, J.J. Steffan, D. Collier, P. Barnhart, and P. Pereira, "Soil and human health: Current status and future needs", *Air. Soil. Water. Res.,* p. 13, 2020.
[http://dx.doi.org/10.1177/1178622120934441]

[18]   M. Vrijheid, M. Casas, M. Gascon, D. Valvi, and M. Nieuwenhuijsen, "Environmental pollutants and child health : A review of recent concerns", *Int. J. Hyg. Environ. Health,* vol. 219, no. 4-5, pp. 331-342, 2016.
[http://dx.doi.org/10.1016/j.ijheh.2016.05.001]

[19]   K.H. Kim, E. Kabir, and S.A. Jahan, "Exposure to pesticides and the associated human health effects", *Sci. Total Environ.,* vol. 575, pp. 525-535, 2017.
[http://dx.doi.org/10.1016/j.scitotenv.2016.09.009]

[20]   "The harmful effects of pesticides on the body. See which are the most dangerous", Available at: https://arhiva.bzi.ro/efectele.nocive-ale-pesticidelor-asupra-organismului-vezi-care-sunt-cele-mai- periculoase-984967

[21]   "Radioactive contamination"., Available at: https://www.nrc.gov/reading-rm/basic-ref/glossary/radioactive-contamination.html

[22]   S. Mittal, A. Rani, R. Mehra, and R.C. Ramola, "Estimation of natural radionuclides in the soil samples and its radiological impact on human health", *Radiat. Eff. Defects Solids,* vol. 173, no. 7-8, pp. 673-682, 2018.
[http://dx.doi.org/10.1080/10420150.2018.1493482]

[23]   O. Komissarova, and T. Paramonova, "Land use in agricultural landscapes with chernozems contaminated after chernobyl accident: Can we be confident in radioecological safety of plant foodstuff?", *Int. Soil Water Conserv. Res.,* vol. 7, no. 2, pp. 158-166, 2019.
[http://dx.doi.org/10.1016/j.iswcr.2019.03.001]

[24]   T.J. Yasunari, A. Stohl, R.S. Hayano, J.F. Burkhart, S. Eckhardt, and T. Yasunari, "Cesium-137 deposition and contamination of Japanese soils due to the Fukushima nuclear accident", *Proc. Natl. Acad. Sci.,* vol. 108, no. 49, pp. 19530-19534, 2011.
[http://dx.doi.org/10.1073/pnas.1112058108]

[25]   C.D.O.B.R.E. Eng, *Optimization of the characterization of contaminated sites in order to remediate them* Technical University of Constructions Bucharest: Bucharest, 2012.

[26]   P. Bardos, ""The contaminated land rehabilitation network for environment technologies in europe", final report for research contract CLL 35/1/12: managing and developing the uk interface with clarinet", Available at: http://www.eugris.info/newsdownloads/final%20report%20clarinet.pdf

[27]   "Technical guide on methods of investigation and assessment of soil and subsoil pollution approved by joint order of the minister of the environment and sustainable development, the minister of economy and finance and the minister of agriculture and rural development", *in accordance with GD no. 1408/2007,* 2007.

[28]   T.E. Butt, H. Akram, C. Mahammedi, and C. House, "Conceptual site model: An intermediary between baseline study and risk assessment", In: *WIT Transactions on Engineering Sciences* vol. 129. WIT Press, 2020.

[29]   Environmental Agency, *Guidance for the Safe Development of Housing on Land Affected by Contamination.* vol. 1. R&D Publication, 2008.

[30]   "Guidelines on conceptual site models, northern territory environment protection authority", *netpa,* 2013.

[31]   "Department of toxic control", Available at: http://dtsc.ca.gov/Publicationforms/upload/Guidance_remediation-Soils.pdf

[32]   "Clarifications to ITRC 2012 ISM-1 Guidance", Available at: https://www.itrcweb.org/ism-1/

[33]   Y. Ming-Ho, *Biological and Health Effects of Pollutants, Environmental Toxicology* 2nd. CRC Press LLC: Boca Raton, USA, 2005.

[34]   A. Keshav Krishna, and K. Rama Mohan, "Distribution, correlation, ecological and health risk assessment of heavy metal contamination in surface soils around an industrial area, Hyderabad, India", *Environ. Earth Sci.,* vol. 75, no. 5, p. 411, 2016.
[http://dx.doi.org/10.1007/s12665-015-5151-7]

[35]   "human health risk assessment", Available at: https://www.epa.gov/risk/human-health-risk-assessment

[36] EXAMPLE EXPOSURE SCENARIOS, "Example exposure scenarios", In: *National Center for Environmental Assessment* U.S. Environmental Protection Agency: Washington.

[37] I. Istrate, D. Cocârță, Z. Wu, and M. Stoian, "Minimizing the health risks from hydrocarbon contaminated soils by using electric field-based treatment for soil remediation", *Sustainability,* vol. 10, no. 1, p. 253, 2018.
[http://dx.doi.org/10.3390/su10010253]

[38] C. Streche, *Experimental research on the treatment of soils contaminated with organic pollutants, by applying thermal and electrical techniques* Bucharest, PhD Thesis, 2014.

[39] S. Crognale, D.M. Cocârță, C. Streche, and A. D'Annibale, "Development of laboratory-scale sequential electrokinetic and biological treatment of chronically hydrocarbon-impacted soils", *New Biotechnology.,* vol. 58, pp. 38-44, 2020.
[http://dx.doi.org/10.1016/j.nbt.2020.04.002]

[40] Y.A. Song, D. Houa, J. B. Zhang, D.A. O'Connor, L. A. Guanghe, G. Qingbao, L.D. Shupeng, and L. Peng, "Environmental and socio-economic sustainability appraisal of contaminated land remediation strategies: A case study at a mega-site in China", *Science of the Total Environment,* vol. 610, pp. 391-401, 2018.
[http://dx.doi.org/10.1016/j.scitotenv.2017.08.016 0048-9697]

[41] "Green and sustainable remediation platform", Available at: https://gsr.epa.gov.tw/gsr_public/EN/Default.aspx

[42] K. Maxwell, B. Kiessling, and J. Buckley, "How clean is clean: A review of the social science of environmental cleanups", *Environ. Res. Lett.,* vol. 13, no. 8, p. 083002, 2018.
[http://dx.doi.org/10.1088/1748-9326/aad74b]

[43] M.J. Kaifer, A. Aguilar, A. Arana, C. Balseiro, I. Torá, J.M. Caleya, and C. Pijls, *Guide to Technologies for the Recovery of Contaminated Soils* Community of Madrid, Ministry of the Environment and Territorial Planning: Madrid, 2004, p. 175.

[44] B. Cao, J. Xu, F. Wang, Y. Zhang, and D. O'Connor, "Vertical barriers for land contamination containment: A review", *Int. J. Environ. Res. Public Health,* vol. 18, no. 23, p. 12643, 2021.
[http://dx.doi.org/10.3390/ijerph182312643]

[45] C.N. Mulligan, R.N. Yong, and B.F. Gibbs, "Remediation technologies for metal-contaminated soils and groundwater: An evaluation", *Eng. Geol.,* vol. 60, no. 1-4, pp. 193-207, 2001.
[http://dx.doi.org/10.1016/S0013-7952(00)00101-0]

[46] K. Aminian, and S. Ameri, "Evaluation of the petroleum technology-based dry soil barrier", *J. Petrol. Sci. Eng.,* vol. 26, no. 1-4, pp. 83-89, 2000.
[http://dx.doi.org/10.1016/S0920-4105(00)00023-1]

[47] FRTR, "Remediation technologies screening matrix and reference guide. version 4.0", Available at: https://frtr.gov/matrix/In-Situ-pH-Control/

[48] P. Song, D. Xu, J. Yue, Y. Ma, S. Dong, and J. Feng, "Recent advances in soil remediation technology for heavy metal contaminated sites: A critical review", *Sci. Total Environ.,* vol. 838, no. 3, p. 156417, 2022.
[http://dx.doi.org/10.1016/j.scitotenv.2022.156417]

[49] C.N. Mulligan, R.N. Yong, and B.F. Gibbs, "Remediation technologies for metal-contaminated soils and groundwater: An evaluation", *Eng. Geol.,* vol. 60, no. 1-4, pp. 193-207, 2001.
[http://dx.doi.org/10.1016/S0013-7952(00)00101-0]

[50] I. Ortiz Bernard, J. Sanz García, M. Dorado Valiño, and S. Villar Fernández, "Techniques for the recovery of contaminated soils", *Depósito legal: M-5,* p. 839, 2007.

[51] S.T. Wait, and D. Thomas, "The characterization of base oil recovered from the low temperature thermal desorption of drill cuttings", *SPE/EPA Exploration and Production Environmental*

Conference, San Antonio, TX, March 10-12, pp.151-158, 2003.

[52]   N. Roca, M. Garcia-Valles, and P. Alfonso, "Fabrication of glass-based products as remediation alternative for contaminated urban soils of Barcelona", *Mater. Lett.,* vol. 305, p. 130741, 2021.
[http://dx.doi.org/10.1016/j.matlet.2021.130741]

[53]   W. Cao, L. Zhang, Y. Miao, and L. Qiu, "Research progress in the enhancement technology of soil vapor extraction of volatile petroleum hydrocarbon pollutants", *Environ. Sci. Process. Impacts,* vol. 23, no. 11, pp. 1650-1662, 2021.
[http://dx.doi.org/10.1039/D1EM00170A]

[54]   Y. Ding, Y. Zhang, Z. Deng, H. Song, J. Wang, and H. Guo, "An innovative method for soil vapor extraction to improve extraction and tail gas treatment efficiency", *Sci. Rep.,* vol. 12, no. 1, p. 6495, 2022.
[http://dx.doi.org/10.1038/s41598-022-08734-8]

[55]   FRTR, "Remediation Technologies Screening Matrix and Reference Guide. Version 4.0", Available at: https://frtr.gov/matrix/Soil-Washing/

[56]   R.W. Peters, "Chelant extraction of heavy metals from contaminated soils", *J. Hazard. Mater.,* vol. 66, no. 1-2, pp. 151-210, 1999.
[http://dx.doi.org/10.1016/S0304-3894(99)00010-2]

[57]   F. Cazals, S. Colombano, D. Huguenot, S. Betelu, N. Galopin, A. Perrault, M-O. Simonnot, I. Ignatiadis, S. Rossano, and M. Crampon, "Polycyclic aromatic hydrocarbons remobilization from contaminated porous media by (bio)surfactants washing", *J. Contam. Hydrol.,* vol. 251, p. 104065, 2022.
[http://dx.doi.org/10.1016/j.jconhyd.2022.104065]

[58]   X.J. Zheng, Q. Li, H. Peng, J.X. Zhang, W.J. Chen, B.C. Zhou, and M. Chen, "Remediation of heavy metal-contaminated soils with soil washing: A review", *Sustainability,* vol. 14, no. 20, p. 13058, 2022.
[http://dx.doi.org/10.3390/su142013058]

[59]   A. Moutsatsou, M. Gregou, D. Matsas, and V. Protonotarios, "Washing as a remediation technology applicable in soils heavily polluted by mining–metallurgical activities", *Chemosphere,* vol. 63, no. 10, pp. 1632-1640, 2006.
[http://dx.doi.org/10.1016/j.chemosphere.2005.10.015]

[60]   A.J. Son, K-H. Shin, J.U. Lee, and K.W. Kim, "Chemical and ecotoxicity assessment of PAH - contaminated soils remediated by enhanced soil flushing", *Environ. Eng. Sci.,* vol. 20, no. 3, pp. 197-206, 2003.
[http://dx.doi.org/10.1089/109287503321671401]

[61]   D. Wen, R. Fu, and Q. Li, "Removal of inorganic contaminants in soil by electrokinetic remediation technologies: A review", *J. Hazard. Mater.,* vol. 401, no. 123345, p. 123345, 2021.
[http://dx.doi.org/10.1016/j.jhazmat.2020.123345]

[62]   R.T. Gill, M.J. Harbottle, J.W.N. Smith, and S.F. Thornton, "Electrokinetic-enhancement bioremediation of organic contaminants: A review of processes and environmental applications", *Chemosphere,* vol. 107, no. 31, 2014.

[63]   Y.B. Acar, and A.N. Alshawabkeh, "Principles of electrokinetic remediation", *Environ. Sci. Technol.,* vol. 27, no. 13, pp. 2638-2647, 1993.
[http://dx.doi.org/10.1021/es00049a002]

[64]   L.Y. Wick, L. Shi, and H. Harms, "Electro-bioremediation of hydrophobic organic soil-contaminants: A review of fundamental interactions", *Electrochim. Acta,* vol. 52, no. 10, pp. 3441-3448, 2007.
[http://dx.doi.org/10.1016/j.electacta.2006.03.117]

[65]   M. Tejada, and J.L. González, "Beet vinasse applied to wheat under dryland conditions affects soil properties and yield", *Eur. J. Agron.,* vol. 23, no. 4, pp. 336-347, 2005.
[http://dx.doi.org/10.1016/j.eja.2005.02.005]

[66] M. Tejada, C. García, J.L. González, and M.T. Hernández, "Use of organic amendment as a strategy for saline soil remediation: Influence on the physical, chemical and biological properties of soil", *Soil Biol. Biochem.,* vol. 38, no. 6, pp. 1413-1421, 2006.
[http://dx.doi.org/10.1016/j.soilbio.2005.10.017]

[67] Y. Liang, Y. Yang, C. Yang, Q. Shen, J. Zhou, and L. Yang, "Soil enzymatic activity and growth of rice and barley as influenced by organic manure in an anthropogenic soil", *Geoderma,* vol. 115, no. 1-2, pp. 149-160, 2003.
[http://dx.doi.org/10.1016/S0016-7061(03)00084-3]

[68] V. Pérez Espinosa, "Immobilization of potentially toxic elements in abandoned mining areas through the construction of technosols and permeable reactive barriers", In: *Thesis defended on June 16.* Universidad de Murcia, 2014.

[69] "US EPA (United States environmental protection agency)", Available at: https://www.epa.gov/remedytech

[70] "US EPA (United States environmental protection agency).contamined site clean-up information", Available at: https://clu-in.org/techfocus/

[71] A. Gusmão, T.M.P. Campos, M.M.M. Nobre, and E.A. Jr, "Laboratory tests for reactive barrier design", *J. Hazard. Mater.,* vol. 110, no. 1-3, pp. 105-112, 2004.
[http://dx.doi.org/10.1016/j.jhazmat.2004.02.043]

[72] R.T. Wilkin, and M.S. McNeil, "Laboratory evaluation of zero-valent iron to treat water impacted by acid mine drainage", *Chemosphere,* vol. 53, no. 7, pp. 715-725, 2003.
[http://dx.doi.org/10.1016/S0045-6535(03)00512-5]

[73] "US EPA (United States environmental protection agency)", Available at: https://www.epa.gov/remedytech

[74] S.V. Ho, C.J. Athmer, P.W. Sheridan, and A.P. Shapiro, "Scale-up aspects of the Lasagna™ process for *in situ* soil decontamination", *J. Hazard. Mater.,* vol. 55, no. 1-3, pp. 39-60, 1997.
[http://dx.doi.org/10.1016/S0304-3894(97)00016-2]

[75] M.L. Benner, R.H. Mohtar, and L.S. Lee, "Factors affecting air sparging remediation systems using field data and numerical simulations", *J. Hazard. Mater.,* vol. 95, no. 3, pp. 305-329, 2002.
[http://dx.doi.org/10.1016/S0304-3894(02)00144-9]

[76] S.F. Kaslusky, and K.S. Udell, "Co-injection of air and steam for the prevention of the downward migration of DNAPLs during steam enhanced extraction: An experimental evaluation of optimum injection ratio predictions", *J. Contam. Hydrol.,* vol. 77, no. 4, pp. 325-347, 2005.
[http://dx.doi.org/10.1016/j.jconhyd.2005.02.003]

[77] FRTR, "Remediation technologies screening matrix and reference guide", Available at: https://frtr.gov/matrix/In-Situ-Chemical-Oxidation/

[78] "US EPA (United States environmental protection agency)", Available at: https://www.epa.gov/remedytech

[79] E. Brillas, J.C. Calpe, and P.L. Cabot, "Degradation of the herbicide 2,4-dichlorophenoxyacetic acid by ozonation catalyzed with Fe2+ and UVA light", *Appl. Catal. B,* vol. 46, no. 2, pp. 381-391, 2003.
[http://dx.doi.org/10.1016/S0926-3373(03)00266-2]

[80] E.R.L. Tiburtius, P. Peralta-Zamora, and A. Emmel, "Treatment of gasoline-contaminated waters by advanced oxidation processes", *J. Hazard. Mater.,* vol. 126, no. 1-3, pp. 86-90, 2005.
[http://dx.doi.org/10.1016/j.jhazmat.2005.06.003]

[81] M. Liu, E. Almatrafi, Y. Zhang, P. Xu, B. Song, C. Zhou, G. Zeng, and Y. Zhu, "A critical review of biochar-based materials for the remediation of heavy metal contaminated environment: Applications and practical evaluations", *Science of The Total Environment,* vol. 806, no. part-1, p. 150531, 2022.
[http://dx.doi.org/10.1016/j.scitotenv.2021.150531]

[82]   FRTR, "Remediation technologies screening matrix and reference guide", Available at: https://frtr.gov/matrix/default.cfm

[83]   FRTR, "Remediation technologies screening matrix and reference guide", Available at: https://frtr.gov/matrix/default.cfm

[84]   V. de Lorenzo, "Systems biology approaches to bioremediation", *Curr. Opin. Biotechnol.,* vol. 19, no. 6, pp. 579-589, 2008.
[http://dx.doi.org/10.1016/j.copbio.2008.10.004]

[85]   FRTR, "Remediation technologies screening matrix and reference guide", Available at: https://frtr.gov/matrix/Phytoremediation/

[86]   P.G. Mihopoulos, M.T. Suidan, and G.D. Sayles, "Complete remediation of PCE contaminated unsaturated soils by sequential anaerobic-aerobic bioventing", *Water Sci. Technol.,* vol. 43, no. 5, pp. 365-372, 2001.
[http://dx.doi.org/10.2166/wst.2001.0325]

[87]   A. Hussain, F. Rehman, H. Rafeeq, M. Waqas, A. Asghar, N. Afsheen, A. Rahdar, M. Bilal, and H.M.N. Iqbal, "*In-situ, Ex-situ,* and nano-remediation strategies to treat polluted soil, water, and air : A review", *Chemosphere,* vol. 289, p. 133252, 2022.
[http://dx.doi.org/10.1016/j.chemosphere.2021.133252]

[88]   B. Lukić, A. Panico, D. Huguenot, M. Fabbricino, E.D. van Hullebusch, and G. Esposito, "A review on the efficiency of landfarming integrated with composting as a soil remediation treatment", *Environ. Technol. Rev.,* vol. 6, no. 1, pp. 94-116, 2017.
[http://dx.doi.org/10.1080/21622515.2017.1310310]

[89]   L. Li, C.J. Cunningham, V. Pas, J.C. Philp, D.A. Barry, and P. Anderson, "Field trial of a new aeration system for enhancing biodegradation in a biopile", *Waste Manag.,* vol. 24, no. 2, pp. 127-137, 2004.
[http://dx.doi.org/10.1016/j.wasman.2003.06.001]

[90]   G. Plaza, K. Ulfig, A. Worsztynowicz, G. Malina, B. Krzeminska, and R.L. Brigmon, "Respirometry for assessing the biodegradation of petroleum hydrocarbons", *Environ. Technol.,* vol. 26, no. 2, pp. 161-170, 2005.
[http://dx.doi.org/10.1080/09593332608618569]

[91]   K.A. Thorn, and K.R. Kennedy, "NMR investigation of the covalent binding of reduced TNT amines to soil humic acid, model compounds, and lignocellulose", *Environ. Sci. Technol.,* vol. 36, no. 17, pp. 3787-3796, 2002.
[http://dx.doi.org/10.1021/es011383j]

[92]   A. Esteve-Núñez, A. Caballero, and J.L. Ramos, "Biological degradation of 2,4,6-Trinitrololuene", *Microbiol. Mol. Biol. Rev.,* vol. 65, no. 3, pp. 335-352, 2001.
[http://dx.doi.org/10.1128/MMBR.65.3.335-352.2001]

[93]   S. Kuppusamy, P. Thavamani, K. Venkateswarlu, Y.B. Lee, R. Naidu, and M. Megharaj, "Remediation approaches for polycyclic aromatic hydrocarbons (PAHs) contaminated soils: Technological constraints, emerging trends and future directions", *Chemosphere,* vol. 168, pp. 944-968, 2017.
[http://dx.doi.org/10.1016/j.chemosphere.2016.10.115]

[94]   A. Parnian, A. Parnian, H. Pirasteh-Anosheh, J.N. Furze, M.N.V. Prasad, M. Race, P. Hulisz, and A. Ferraro, "Full-scale bioremediation of petroleum-contaminated soils *via* integration of co-composting", *J. Soils Sediments,* vol. 22, no. 8, pp. 2209-2218, 2022.
[http://dx.doi.org/10.1007/s11368-022-03229-5]

[95]   M.A. Rao, G. Di Rauso Simeone, R. Scelza, and P. Conte, "Biochar based remediation of water and soil contaminated by phenanthrene and pentachlorophenol", *Chemosphere,* vol. 186, pp. 193-201, 2017.
[http://dx.doi.org/10.1016/j.chemosphere.2017.07.125]

[96]   J.M. Castillo Diaz, L. Delgado-Moreno, R. Núñez, R. Nogales, and E. Romero, "Enhancing pesticide degradation using indigenous microorganisms isolated under high pesticide load in bioremediation systems with vermicomposts", *Bioresour. Technol.,* vol. 214, pp. 234-241, 2016.
[http://dx.doi.org/10.1016/j.biortech.2016.04.105]

[97]   S. Smith, "A critical review of the bioavailability and impacts of heavy metals in municipal solid waste composts compared to sewage sludge", *Environ. Int.,* vol. 35, no. 1, pp. 142-156, 2009.
[http://dx.doi.org/10.1016/j.envint.2008.06.009]

[98]   Y.D. Motasem Alazaiza, A. Albahnasawi, N.K. Copty, A.M. Gomaa Ali, J.K. Mohammed, M. Tahra Al, S. Salem, Amr Abu, S. Mohammed Shadi, J. Abu, and N. Dia Eddin, "Thermal based remediation technologies for soil and groundwater: A review", *Desal.Water. Treat.,* vol. 259, pp. 206-220, 2022.

[99]   J. Piña, J. Merino, A.F. Errazu, and V. Bucalá, "Thermal treatment of soils contaminated with gas oil: influence of soil composition and treatment temperature", *J. Hazard. Mater.,* vol. 94, no. 3, pp. 273-290, 2002.
[http://dx.doi.org/10.1016/S0304-3894(02)00081-X]

[100] "FRTR", "Remediation technologies screening matrix and reference guide", Available at: https://frtr.gov/matrix/Desorption-Incineration/

[101] C. Zhao, Y. Dong, Y. Feng, Y. Li, and Y. Dong, "Thermal desorption for remediation of contaminated soil: A review", *Chemosphere,* vol. 221, pp. 841-855, 2019.
[http://dx.doi.org/10.1016/j.chemosphere.2019.01.079]

CHAPTER 5

Risk–based Approach to Air Quality Management

Marius D. Bontoş[1,*]

[1] *Department of Hydraulics, Hydraulic Machinery and Environmental Engineering, University POLITEHNICA of Bucharest, Faculty of Energy Engineering, Bucharest, Romania*

Abstract: In the last decades, increased air pollution has been the world's largest environmental health threat, and new causal relationships between it and human diseases have been discovered. To better understand the relationship between air pollution and health risks and to promote the most efficient measures that may reduce the health impact, the chapter sets the context regarding the risk–based approach to air quality management and presents the Health Impact Assessment of Air Pollution, the Health Risk Assessment process and the tools that can be used for assessing it. To support the theoretical information described, several case studies were presented.

Keywords: Air pollution, Health impact assessment, Health risk assessment, Hazard identification, Exposure assessment, Dose-response assessment, Risk characterization, Air quality index, Years of life lost, Disability-adjusted life years, Change in life expectancy, Geographic information system.

INTRODUCTION

Air pollution is defined by the World Health Organisation as the "*contamination of the indoor or outdoor environment by any chemical, physical or biological agent that modifies the natural characteristics of the atmosphere*" [1]. Gases (like carbon, sulphur and nitrous oxides, ammonia, methane, and CFCs), particulates (inorganic and organic), and viruses and bacteria are the common pollutants of the air. At the global level, the climate and ecosystems are closely related to air quality. Fossil fuel burning is the main cause of air pollution and contributes to greenhouse gas emissions.

At the local scale, the connections between climate and air pollution are also very important, as they influence each other. The rapid and large-amplitude variations of pollutant concentrations are mostly generated by a series of meteorological phenomena that occur on this scale, such as the thermal inversion layer, sea and

---

* **Corresponding author Marius D. Bontoş:** Department of Hydraulics, Hydraulic Machinery and Environmental Engineering, University POLITEHNICA of Bucharest, Faculty of Energy Engineering, Romania; E-mail: bontos.marius@upb.ro

valley breezes, the canyon effect, *etc*. The local scale pollution primarily affects population health through direct short-term action, but also through longer-term toxicity for certain pathologies.

According to United Nations Economic Commission for Europe (UNECE) and World Health Organization (WHO), air pollution is considered to be *"the world's largest environmental health threat"*. The WHO reports that almost 99% of the global population breathe polluted air that exceeds currently set guidelines; population from countries with low and middle income being the ones that suffer from the highest exposures [1]. Also, over 90% of residents in the European region are exposed to yearly levels of outdoor fine particulate matter that exceed WHO and European Environment Agency (EEA) air quality limits, meaning that almost every single person is affected by air pollution [2].

Air pollution from both indoor and outdoor sources contributes significantly to morbidity and mortality by causing allergies, respiratory diseases, and other disorders with both short-term and long-term effects. The effects of air pollution differ from one group of people to another, some individuals, such as children and elderly ones, being more sensitive to pollutants. Also, people with known health problems such as heart and lung disease or asthma may experience exacerbations of these conditions when the air is polluted. The degree to which a person is affected by air pollution is dependent on the exposure time and concentration of harmful substances.

Currently, considering the increased number of experimental and epidemiological studies performed in the last decades on different areas and populations, the health risks related to air pollution are established more accurately, especially for short-term effects. These studies highlight the role of air pollution in the occurrence or exacerbation of a wide range of health effects that can vary from cardio-respiratory diseases to early mortality.

According to data provided by WHO [1], each year, polluted air is responsible for 1.4 million (24%) of all deaths from stroke, 1.8 million (43%) of all deaths from lung disease and lung cancer, as well as for 2.4 million (25%) of all heart disease deaths worldwide.

European environmental protection policy has taken into account the impact on health from the beginning. The leading European and International environmental organizations have stated that, although many environmental and health problems have been ameliorated or solved, important steps remain to be taken, especially regarding the health effects of chronic exposures [3, 4].

The relationships between the environment and health have turned out to be much more complicated than previously thought, being marked by many causal links. In other words, the relationships between exposures and health effects depend on the pollutants present in the air but are also influenced by different factors such as genetic makeup, age, nutrition, lifestyle, and socioeconomic ones.

Due to the increased number of deaths that can be related to air pollution, it is necessary to implement health impact assessments at local scale and to translate the results into local, regional or national policies with the aim of reducing the impact on human health.

HEALTH IMPACT ASSESSMENT OF AIR POLLUTION

Health Impact Assessment (HIA) provides useful information to decision makers about how a policy, programme or project may affect people's health. Because of its ability to influence decision makers and stakeholders, WHO promotes the use of HIA.

HIA has been defined by many different people and organizations. All definitions are almost similar, differing through the emphasis given to particular components of the HIA approach. Some of the most comprehensive definitions are presented below:

o World Health Organization defines Health Impact Assessment, that can also be implemented for air pollution, as a *"practical approach used to judge the potential health effects of a policy, programme or project on a population, particularly on vulnerable or disadvantaged groups"* [5]. To maximize the proposal's beneficial health effects and minimize its negative health impact, recommendations must be made for decision-makers and stakeholders.

o European Centre for Health Policy (ECHP) defines HIA as *"a combination of procedures, methods, and tools by which a policy, program, or project may be judged as to its potential effects on the health of a population, and the distribution of those effects within the population"* [6].

o According to the United States Environmental Protection Agency (EPA), HIA is *"a tool designed to investigate how a proposed program, project, policy, or plan may impact health and well-being and inform decision-makers of these potential outcomes before the decision is made"* [7].

Health Impact Assessment process consists of 5 steps (Fig. **5.1**):

o **Screening**: an intervention, a policy, or a project for which a health impact assessment would be beneficial is selected. There are three possible outcomes based on the potential consequences of health determinants, health outcomes, and population groups:

Fig. (5.1). Health Impact Assessment process steps.

• Health Impact Assessment is required

• Health Impact Assessment is unnecessary because the effects are previously known

• Health Impact Assessment is unnecessary since the effects are insignificant.

o **Scoping**: planning HIA activities and identifying what health risks and benefits to consider. This step defines, among others, which elements or aspects of the proposal will be assessed, the aims and objectives of the assessment, timescale, the geographical area covered, the populations/communities affected by the implementation, and any vulnerable, marginalised, or disadvantaged groups.

o **Appraisal/Health Risk Assessment**: the core of any HIA activity. All the data and evidence are collected and analysed, affected populations are identified, and health impacts are estimated giving suggestions and recommendations for actions that promote positive health effects and minimize the negative ones.

o **Reporting**: the contents of the report should include a description of the scope, the priorities identified at the beginning of the process, the opinions expressed by the stakeholders, the evidence available from the various sources, the overall findings, and any other recommendations.

o **Monitoring**: represents HIA's final step which allows process evaluation and its effectiveness.

AIR POLLUTION HEALTH RISK ASSESSMENT

Health risk assessment (HRA) of air pollution is "*the scientific evaluation of potential adverse health effects resulting from human exposure to a particular hazard*", being the core of the Health Impact Assessment process [8].

To provide more precise estimations, the HRA combines studies results of the health effects at different exposure levels with the outcomes provided by the studies that evaluate the exposure level at various distances from the pollutant source.

Scientists and decision-makers use a four-step risk assessment process (Fig. **5.2**) that includes [9]:

o Hazard Identification

o Exposure assessment

o Dose-Response Assessment

o Risk characterization

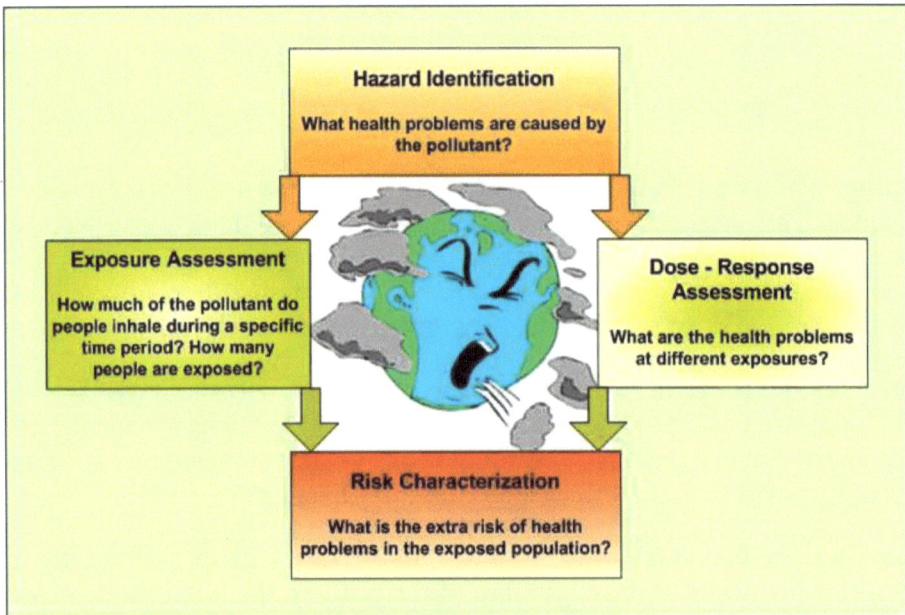

Fig. (5.2). The four-step risk assessment process (adapted from EPA [9]).

Hazard Identification

Identifying hazards involves determining whether the substance may or may not harm humans. Often, the problem identification process begins with scientific or public concerns about a particular substance. Human harm rarely has direct evidence, instead, evidence is obtained through:

o Animal studies. If the tested substance induces health problems in animals, it is considered that it can also affect human health too.

o *In vitro* studies. Experiments on cells or microorganisms are used to examine how chemicals can produce toxic effects.

o Comparative studies. The characteristics of the studied substance are compared with the characteristics of substances with known harmful effects.

o Epidemiological studies. Even though they provide the most reliable results, only those that are relevant to the studied population should be performed.

The most dangerous air pollutants are those that have a significant negative impact on public health or that affect a large part of the population. The related health problems can include cardiovascular diseases, cancer, respiratory diseases, allergies, neurological and psychological effects, disruption of the endocrine system and metabolism and birth defects. The correlations between the most dangerous air pollutants and their health effects can be seen in (Fig. **5.3**).

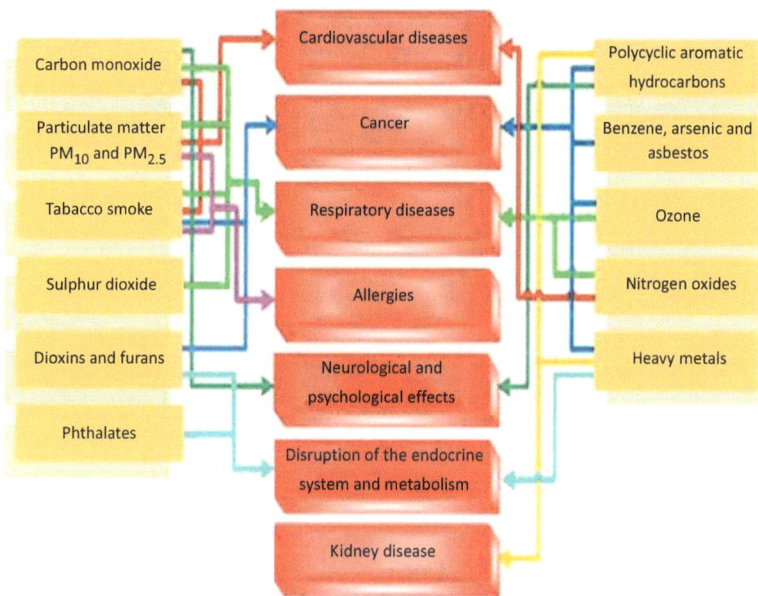

Fig. (5.3). The relationship between air pollutants and their effects on human health.

Although we are all exposed to the effects of air pollution, certain groups of people are more susceptible. Individual reactions to airborne contaminants depend on several factors, such as the type of pollutant, level of exposure, age, and health.

Using the air pollution pyramid (Fig. **5.4**), one can determine the relative order and severity of health impacts by tracking the proportion of the population affected.

The air pollution health effects pyramid is "*a diagrammatic presentation of the relationship between the severity and frequency of health effects, with the mildest and most common effects at the bottom of the pyramid, e.g.*, symptoms, and the least common but more severe at the top of the pyramid, *e.g.*, premature mortality. The pyramid demonstrates that as severity decreases, the number of people affected increases" [10].

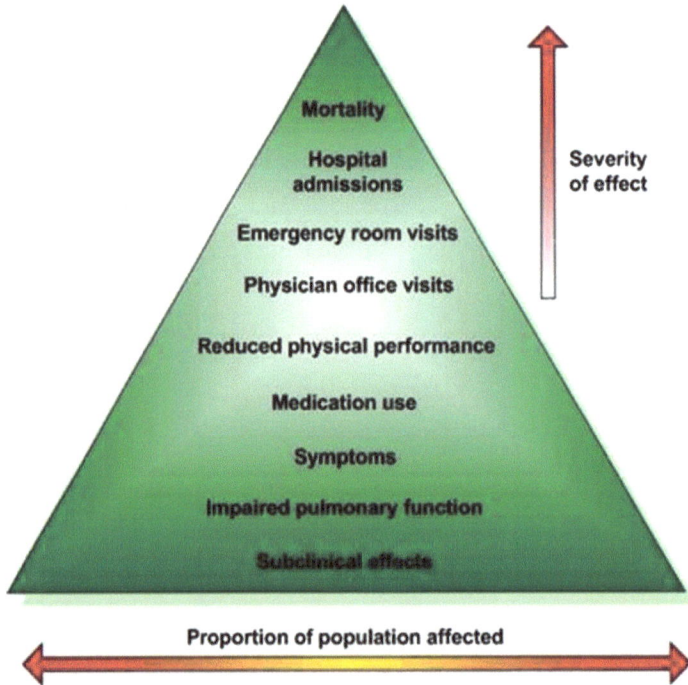

Fig. (5.4). Air pollution health effects pyramid [10].

Exposure Assessment

All gases or particles present in the atmosphere, that can have harmful effects on human health or on the environment, are considered to be air pollutants. They can be classified as primary or secondary based on their formation. Primary pollutants are those released directly into the atmosphere by pollution sources, while secondary pollutants may occur due to reactions between primary ones. An example of both a primary and secondary pollutant is $PM_{2.5}$ (particles smaller than 2.5 μm). These particles appear from combustion processes but may also be formed by reactions between nitrogen oxides (NO_x), volatile organic compounds (VOCs), and sulphur oxides (SO_x). An example of a secondary pollutant is ground level ozone, which is formed through reactions between nitrogen oxides and volatile organic compounds in the presence of sunlight [11].

A list of the main pollutants that can have negative effects on human health, together with their description and information related to the major sources of pollution associated, can be seen in Table **5.1** [11, 12].

Table 5.1. The list of most important air pollutants.

Pollutant	Description	Major sources
Particulate matter (PM)	• Mixture of solid and liquid particles with different chemical and physical properties in air • PM_{10} particles are 10 microns (μm) in aerodynamic diameter and smaller • $PM_{2.5}$ particles are 2.5 μm in aerodynamic diameter and smaller; $PM_{2.5}$ particles can penetrate further down in the respiratory system compared to larger particles	• Motor vehicle engines, industrial processes, wood burning; breakdown of materials including earth's crust; reactions between pollutants such as NO_x, VOCs and NH_3
Ozone	• A reactive oxygen species	• Reactions between NO_x and VOCs in presence of sunlight
Nitrogen oxides (NO_x)	• A group of reactive gases that include nitric oxide (NO) and nitrogen dioxide (NO_2) • NO_2 is odorous, brown and highly corrosive	• Motor vehicles, wood burning, industrial processes (power generation, use of industrial boilers and diesel generators, petroleum refining)
Sulphur dioxide (SO_2)	• A colourless gas that has a pungent odour that smells like a struck match.	• Marine vessels, smelting, petroleum refining, diesel engines
Carbon monoxide	• A colourless, odourless gas produced from incomplete combustion of fuel.	• Motor vehicles, waste incineration space heating
Volatile organic compounds	• A group of carbon-containing gases and vapors (e.g., benzene, toluene, xylene)	• Transportation, industry (oil and gas, petroleum refining, pulp and paper mills), consumer products (solvents, paints, cleaning products), residential wood combustion • React with NO_x to form ozone
Ammonia	• A colourless gas with a pungent odor	• Agricultural activities

The amount of pollution, its severity, frequency and duration, as well as the number of persons who are exposed, are all determined through exposure assessment [9].

Exposure may be long-term or short-term, occupational, or environmental. There are three primary routes of exposure to chemical contaminants: dermal (skin-absorbed), respiratory (inhaled), and gastrointestinal (ingested). Exposure can be assessed in two different ways: directly, through tests on body fluids or tissues, or indirectly, through analysis of the contaminant levels in the environment.

Using indirect estimation, the exposure assessment is also a four-step process (Fig. **5.5**):

- Identification of pollutants that are likely to be present in the air.

- Estimation of the quantities of these pollutants discharged from various sources.

- Estimation of pollutant concentrations for the geographic areas of interest.

- Estimation of the total number of people breathing polluted air at different levels or at a predetermined level, such as a regulatory standard or a health reference level.

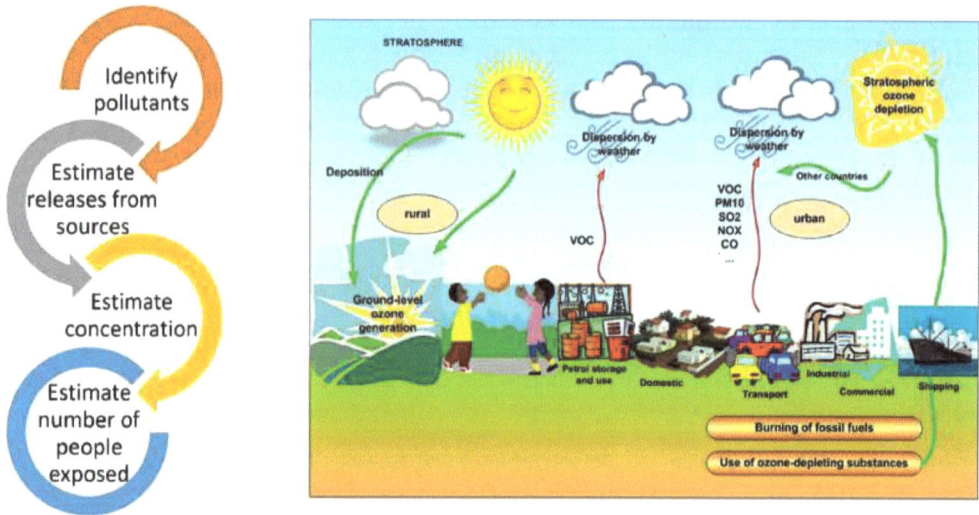

Fig. (5.5). Exposure assessment main steps.

Dose-Response Assessment

"All substances are poisons: there is none which is not a poison. The right dose differentiates a poison and a remedy." Paracelsus (1493-1541)

There is no single measure of toxicity. The dose-response concept underlies all toxicity assessments, and it is used differently to assess acute and chronic effects.

Dose-response assessment is the process of determining the relation between the dose of an identified pollutant and the frequency of a negative health effect in populations exposed to that pollutant. It also estimates the incidence of the response associated with human exposure to the considered pollutant. The term "dose" is frequently used to describe the quantity of the pollutant, whereas the term "response" describes the outcome of the pollutant exposure [13, 14].

In other words, dose-response assessment relies on both qualitative and quantitative toxicity information to estimate the dose-response relationship. It describes the link between exposure to a particular pollutant and observed health impact and estimates how different levels of exposure induce changes in the occurrence and severity of health effects.

The dose-response relationship is determined using data obtained through experimental studies on animals, clinical studies on the affected population, or cell studies. While constant doses can only be used in animal studies, long-term human exposures are variable. This can be a significant source of uncertainty and an integrated long-term exposure estimate needs to be developed [9].

Dose-response relationships are determined graphically (dose-response curve, (Fig. **5.6**) by establishing the negative effect generated by the variation of the administered dose. In general, increasing the dose will lead to a proportional increase in both the incidence and severity of the adverse effect.

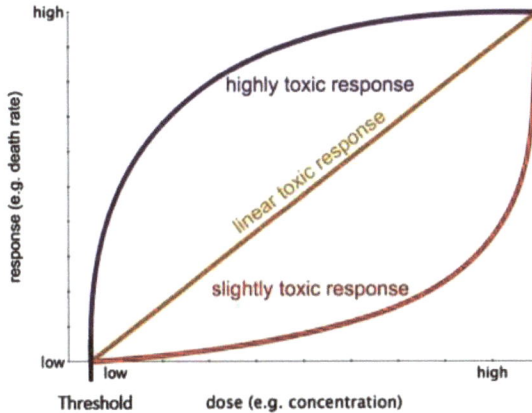

Fig. (5.6). The dose-response curve.

A threshold is an important concept in dose-response relationships. It is generally assumed that there is a minimum dose (threshold), below which exposure to the chemical does not result in adverse effects [9, 13, 14].

Acute responses such as death allow thresholds to be easily determined, while thresholds for chemicals that cause cancer or other chronic diseases are much more difficult to establish. Even in these cases, it is important for toxicologists to identify a specific level of exposure to a chemical for which there is no adverse effect.

Risk Characterisation

The final step of the health risk assessment process is risk characterisation. It integrates all three previous phases (Fig. **5.7**). This phase determines the probability of a toxic substance having a negative effect on a human population by quantifying the number of air pollution-related premature deaths, disability-adjusted life years, and disease cases, as well as defining permissible exposure levels from which exposure standards are established.

Considering the different types of health effects (carcinogenic or non-carcinogenic), risk characterisation data can be calculated and presented in different ways to show how populations or individuals may be affected.

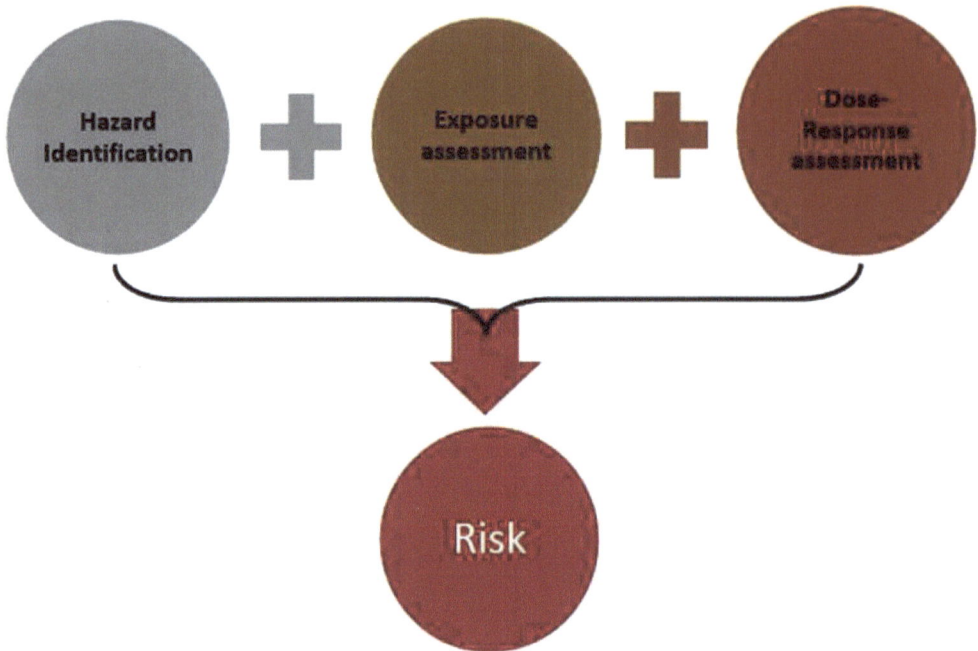

Fig. (5.7). Risk characterisation.

For non-cancer risks, health reference levels or an air quality index is used by the governmental agencies to characterize possible health effects. These levels are developed based on exposure thresholds, resulting from experimental studies, that do not induce health effects.

For example, in a European area, pollutant concentrations limits that shall not be exceeded in a given period of time are set by the EU's air quality directives [15, 16]. Starting from these directives, in the event of exceedances, each national authority must establish and implement air quality management plans to reduce air pollution concentrations levels below the target values.

Table **5.2** summarizes selected EU standards as well as World Health Organization guidelines. Because the reported health effects linked with the various pollutants occur over different exposure times, these standards are applied also across different time periods. The WHO guideline values established to protect human health are often stricter than corresponding politically agreed EU standards [17].

Table 5.2. EU air quality directives and WHO guidelines comparation.

Pollutant	Averaging period	EU Air Quality Directives Objective and legal nature and concentration	Comments	WHO Guidelines Concentration	Comments
$PM_{2.5}$	Daily			$25\mu g/m^3$	99th percentile (3 days per year)
$PM_{2.5}$	Annual	Limit value: $25\mu g/m^3$		$10\mu g/m^3$	
$PM_{2.5}$	Annual	Indicative limit value, $20\mu g/m^3$			
PM_{10}	Daily	Limit value: $50\mu g/m^3$	Not to be exceeded on more than 35 days per year	$50\mu g/m^3$	99th percentile (3 days per year)
PM_{10}	Annual	Limit value: $40\mu g/m^3$		$20\mu g/m^3$	
O_3	Maximum daily 8-hour mean	Target value: $120\mu g/m^3$	Not to be exceeded on more than 25 days per year, averaged over 3 years	$100\mu g/m^3$	
O_3	Maximum daily 8-hour mean	Long term objective: $120\mu g/m^3$			
NO_2	Hourly	Limit value: $200\mu g/m^3$	Not to be exceeded on more than 18 hours per year	$200\mu g/m^3$	
NO_2	Annual	Limit value: $40\mu g/m^3$		$40\mu g/m^3$	
SO_2	Hourly	Limit value: $350\mu g/m^3$	Not to be exceeded on more than 24 hours per year		
SO_2	Daily	Limit value: $125\mu g/m^3$	Not to be exceeded on more than 3 days per year	$20\mu g/m^3$	
CO	Maximum daily 8-hour mean	Limit value: $10\mu g/m^3$		$10\mu g/m^3$	
C_6H_6	Annual	Limit value: $5\mu g/m^3$		$1.7\mu g/m^3$	Reference level
Pb	Annual	Limit value: $0.5\mu g/m^3$	Measured as content in PM_{10}	$0.5\mu g/m^3$	
As	Annual	Limit value: $6ng/m^3$	Measured as content in PM_{10}	$6.6ng/m^3$	Reference level
Cd	Annual	Limit value: $5ng/m^3$	Measured as content in PM_{10}	$5ng/m^3$	
Ni	Annual	Limit value: $20ng/m^3$	Measured as content in PM_{10}	$25ng/m^3$	Reference level

Another way to characterize air quality levels and the possible health effects in Europe is the European Air Quality Index. The European Environment Agency's portal provides access to up-to-date Indexes for each EEA member countries, interested parties being able to obtain information on air quality in individual countries, regions, and cities.

The Index presented in Table **5.3** is calculated hourly for more than two thousand air quality monitoring stations across Europe, taking into account the concentration values for five key pollutants: particulate matter (PM_{10}); fine particulate matter ($PM_{2.5}$); ozone (O_3); nitrogen dioxide (NO_2) and sulphur dioxide (SO_2). It shows the potential impact of air quality on health, determined by the pollutant for which the concentrations are the highest [18].

Table 5.3. The European Air Quality Index.

Pollutant	Index level (based on pollutant concentrations in µg/m³)					
	Good	Fair	Moderate	Poor	Very poor	Extremely poor
Particles less than 2.5 µm ($PM_{2.5}$)	0-10	10-20	20-25	25-50	50-75	75-800
Particles less than 10 µm (PM_{10})	0-20	20-40	40-50	50-100	100-150	150-1200
Nitrogen dioxide (NO_2)	0-40	40-90	90-120	120-230	230-340	340-1000
Ozone (O_3)	0-50	50-100	100-130	130-240	240-380	380-800
Sulphur dioxide (SO_2)	0-100	100-200	200-350	350-500	500-750	750-1250

Source: EU Air Quality Directives (2008/50/EC, 2004/107/EC), WHO, 2006, Air quality guidelines: Global update 2005.

Several metrics are frequently used to sum up the outcomes of the risk characterization step, such as the number of deaths or diseases (ED), years of life lost (YLL), disability-adjusted life years (DALY), or change in life expectancy [19].

The following formula can be used to determine the excess deaths or diseases (ED) caused by an increase in pollutant concentration:

$$ED = \frac{p(RR-1)}{p(RR-1)+1} * I * P \tag{5.1}$$

where *RR* is the relative risk of premature mortality, *p* represents the fraction of the population exposed, *I* is the mortality incidence per year, and *P* is the all-age population.

The number of years of life lost, or *YLL*, is another metric of risk characterisation step. The fundamental formula is:

$$YLL = N * L \tag{5.2}$$

where *N* is the number of deaths, and *L* represents standard life expectancy at the age of death in years.

The basic formula for years lost due to disability (*YLD*) is shown below:

$$YLD = I * DW * L \tag{5.3}$$

where *I* is the number of cases, *L* the average years of disease, and *DW* is the disability weight ranging from 0 (healthy) to 1 (dead).

The Disability-Adjusted Life Year (DALY) for a population group can be considered as the difference between an ideal health status where all people have no disease and disability and the current health status. DALYs can be calculated as the sum of YLL and YLD:

$$DALY = YLL + YLD \tag{5.4}$$

Air Pollution Risk Assessment for Europe (Case Study)

Considering the assessment steps presented above, a short example of a health risk assessment carried out at the European level (for EU-28, Romania, Spain and Italy) is presented below.

• Pollutant identification, emission estimation and sources distribution

The European Environment Agency's interactive database provides access to data from the EU Emissions Inventory Report 1990-2019 under the UNECE Convention on Long-range Transboundary Air Pollution (LRTAP). Based on these data, the distribution graphs of pollutant emissions for EU-28, Romania, Spain, and Italy were made [20].

As can be seen in (Fig. **5.8**), the distribution of emissions in the 4 graphs is similar, Romania has a higher proportion of sulphur oxides (SO$_x$) pollution, of about 56% of total emissions, compared to EU 28, Spain, and Italy. The Gothenburg Protocol Ceilings has been exceeded only by Spain for ammonia (NH$_3$) pollutant.

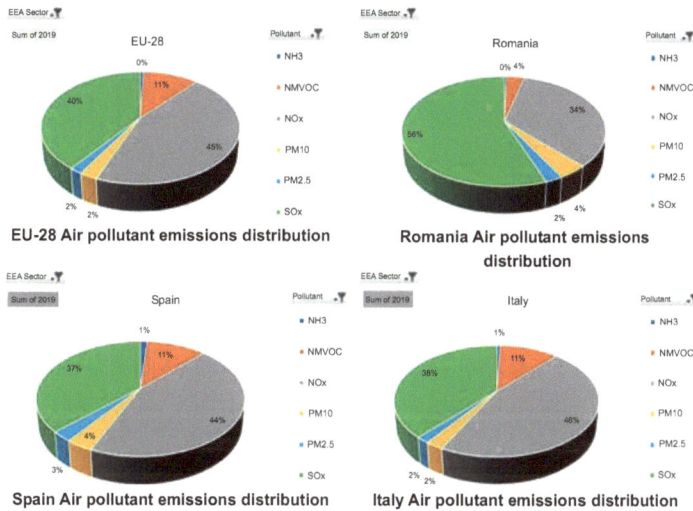

Fig. (5.8). Air pollutant emission distribution.

Also, the European Environment Agency air pollutant emissions data were used to evidence the air pollutant sources distribution for EU-28, Romania, Spain and Italy [20]. As can be seen in Fig. (**5.9**), the main pollution source in EU-28, Spain, and Italy in 2019 was the agricultural sector and in Romania, residential, commercial, and institutional sectors.

• Air pollution risk assessment results

Even today, the high concentrations of air pollutants have a major impact on the health of European citizens, particulate matter (PM$_{2.5}$), nitrogen dioxide (NO$_2$) and ground-level ozone (O$_3$), causing the most health problems.

As can be seen from collected data from the 2020 EEA report Table **5.4**, $PM_{2.5}$ pollution was responsible for an estimated 275 000 premature deaths in all EU countries. Also, in Romania, Italy, and Spain $PM_{2.5}$ concentrations were responsible for the highest number of premature deaths compared to NO_2 and O_3 concentrations [21].

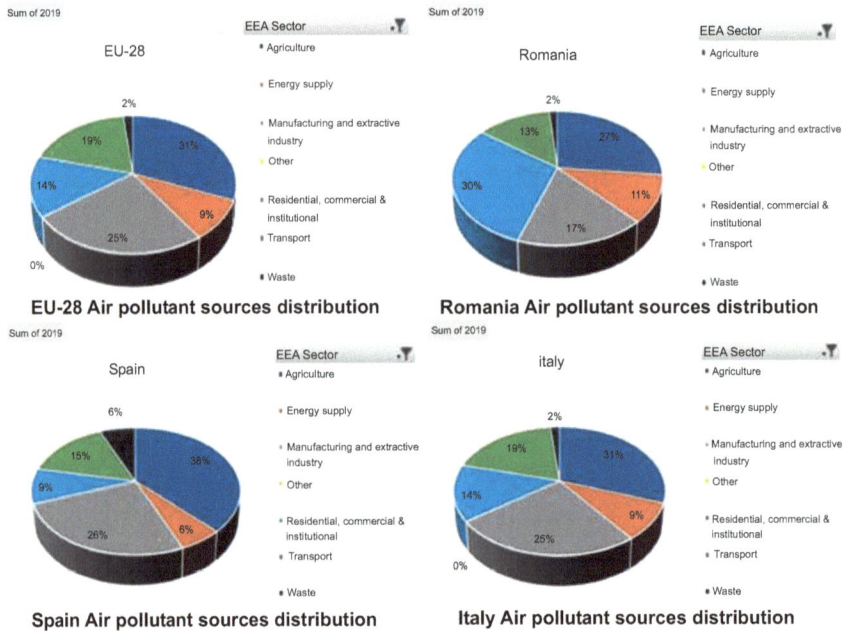

Fig. (5.9). Distribution of air pollution sources.

Table 5.4. Premature deaths related to $PM_{2.5}$, NO_2 and O_3 pollution in 2020 [21].

Country	Population (1000)	$PM_{2.5}$		NO_2		O_3	
		Annual mean	Premature deaths	Annual mean	Premature deaths	SOMO 35	Premature deaths
Italy	59641	15	52300	17.7	11200	6067	5100
Romania	19329	15.2	21600	15.1	3100	2955	1000
Spain	45166	10	17000	14.6	4800	4522	2400
EU-27 total	442851	11.2	238000	14.1	49000	4182	24000
All EU countries	475456	11.4	275000	15.7	64000	4228	28000

Although the approximations provided by these assessments are not perfect, they still provide useful information for scientists to assess the risks associated with air pollution. Also, based on these risk assessments, policymakers can set new regulatory standards to reduce exposure to toxic air pollutants and to minimize the risk of related health problems.

TOOLS FOR ASSESSING HEALTH RISKS FROM AIR POLLUTION

Health Impact/Risk Assessment Tools

There are a variety of tools available that can be useful in assisting public health practitioners and decision makers in their efforts to enhance local air quality and reduce the health effects of pollution. These health impact or risk assessment tools were created by governmental and non-governmental organizations to provide up to date information about air pollutant exposures and their health effects [19].

Most used ones are presented and compared in Table **5.5**. All of them can be used to determine the number of air pollution-related premature deaths, disability-adjusted life years, and disease cases.

Although these tools may use common data sources, they differ in terms of usage, technical complexity, and output format. As can be seen from Table **5.5**, most of these tools can estimate the effects of Nitrogen oxides (NOx), Sulphur Oxides (SOx), and Particulate Matters (PM_x).

Table 5.5. The list of the most used health risk assessment tools and their comparison [19].

Tool	Developer	Pollutants	Risk Assessment
AirQ+: software tool for health risk assessment of air pollution	World Health Organization	$PM_{2.5}$, PM_{10}, O_3, NO_2, SO_2, CO	Mortality, Morbidity, DALY
EIS-PA: assessment of the health impact of urban atmospheric pollution	French Institute of Public Health Surveillance	$PM_{2.5}$, PM_{10}, O_3, NO_2, SO_2, CO, FN	Mortality, Morbidity
BenMap-CE: Environmental Benefits Mapping and Analysis Program - Community Edition	The United States Environmental Protection Agency	$PM_{2.5}$, PM_{10}, O_3, NO_2, SO_2, CO	Mortality, Morbidity, DALY
COBRA: CO-Benefits Risk Assessment Health Impacts Screening and Mapping Tool	The United States Environmental Protection Agency	$PM_{2.5}$, NO_2, SO_2	Mortality, Morbidity, DALY
SIM-Air: The Simple Interactive Model for better Air quality	Urban Emission	$PM_{2.5}$, PM_{10}	Mortality, Morbidity

EIS-PA: Short-term Health Risk Assessment Tool (Case Study)

The French Institute for Health Surveillance (InVS) developed, in 1997, an epidemiological surveillance model to quantify the relationships between pollution present in urban areas and its short-term health effects. This model has been constantly improved, being implemented in over 26 cities in 12 European countries [22]. Starting from this model and using data acquisition, processing and validation, relational database management systems, virtual instrumentation, and geographic information systems (GIS), we have developed an interactive tool that can be used to assess the health risks of air pollution [23].

The tool interface is divided into two different tabs: one for air pollution and health data input (Fig. **5.10**) and the other one for health impact assessment results (Fig. **5.11**).

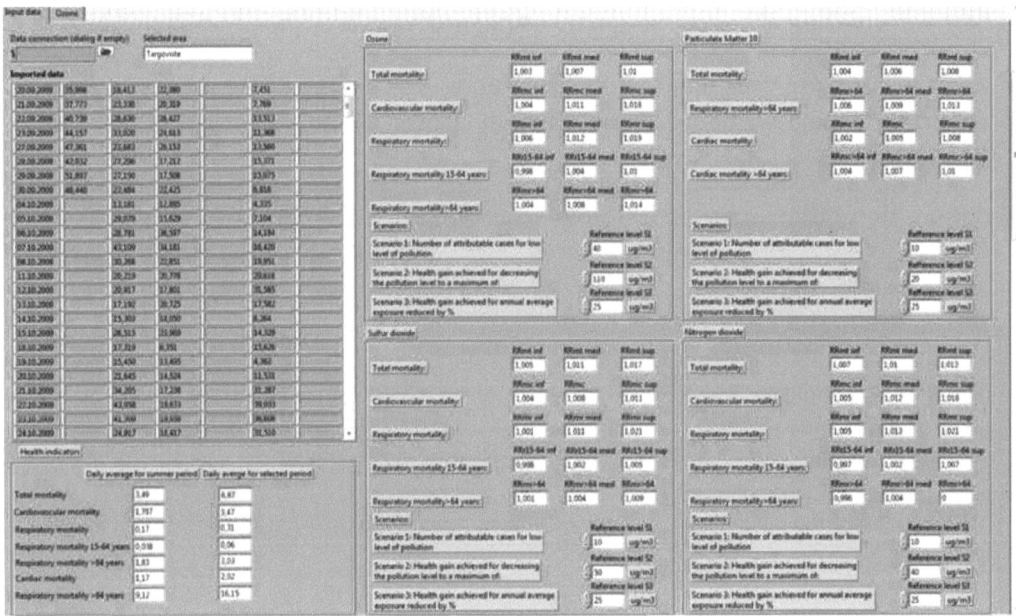

Fig. (5.10). Input data window of health risk assessment tool.

This tool can be used for any urban area where people's exposure to pollutants present in the air can be evaluated accurately and considered homogeneous. To fulfil the hypothesis of homogeneous air quality, only urban areas, including the central part of the city, must be selected. Peripheral areas separated from the town by a green belt or by water and populated areas located near the major industrial pollution sources must be excluded.

For the selected area, the presence of a monitoring network that reliably provides air quality measurements for at least twelve months is required. The air pollution input data can be stored in an Excel file or in a database and must contain the hourly averages provided by surveillance stations for at least twelve months. If the monitoring period is not continuous, it will be extended by a number of days equal to the number of missing days.

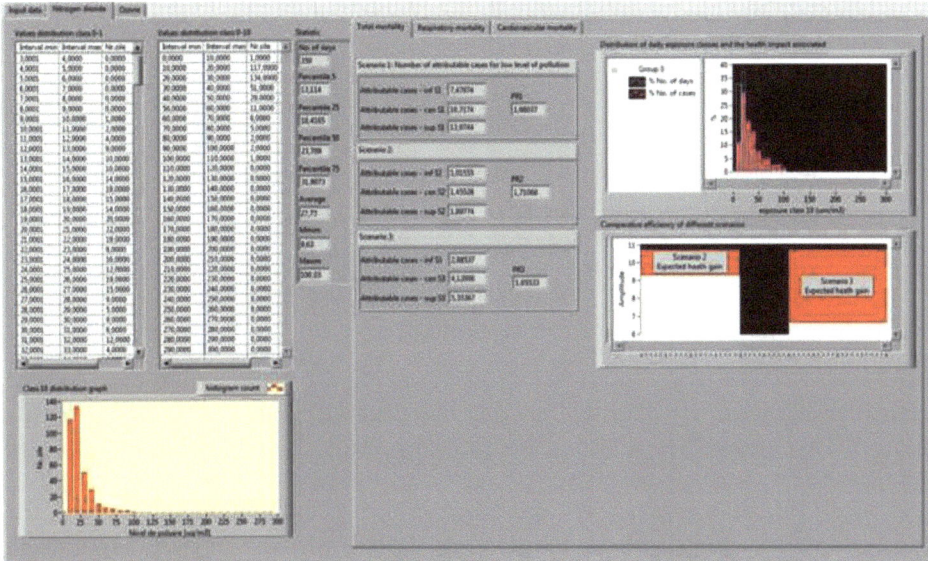

Fig. (5.11). Health risk assessment results interface.

The other input data required for the assessment refers to the total number of people living or working in the selected area, and health data (daily mortality and morbidity), classified according to the World Health Organization regulations, provided by the public health departments for the same area.

After all input data are provided, the short-term Health Risk Assessment tool will automatically calculate the daily exposure indicators and the daily and seasonal assessment for each pollutant.

The daily exposure indicators are calculated as the arithmetic mean of the daily measurements, and their distribution is presented as a histogram using a 10 µg/m^3 exposure step (Fig. **5.12**).

The daily and seasonal assessment for each pollutant will consider three scenarios:

- The health gain for exposure to low levels of pollution (*e.g.*, 10µg/m^3).

- The health gain attributable to suppression of pollution levels to EU air quality limits.

- The health gain achieved for annual average exposure reduced by 25%.

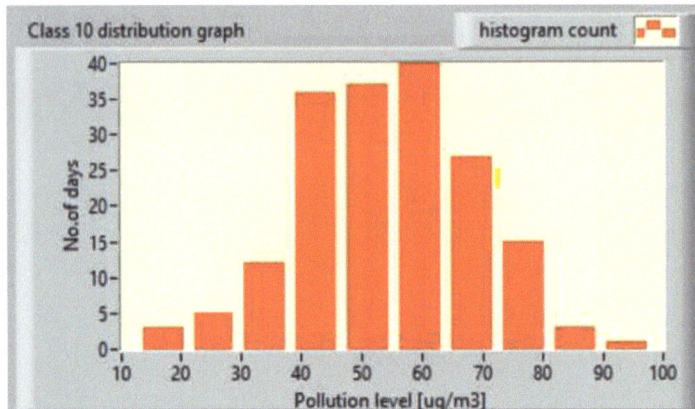

Fig. (5.12). The exposure indicators distribution histogram.

The number of sanitary events that can be attributed to urban air pollution is calculated for each of the exposure indicators and for each day of the studied period.

An example of the graphical distribution of daily exposure classes and the health impact associated with each class can be seen in Fig. (**5.13**).

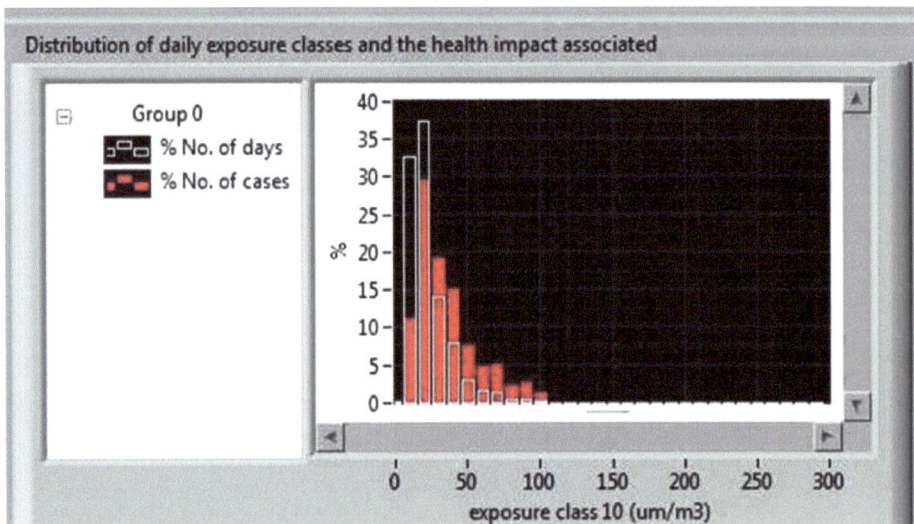

Fig. (5.13). Graphical distribution of daily exposure classes and the associated health impact.

The seasonal sanitary impact is obtained by summing the sanitary events calculated for each day of the study period and for each scenario. An example of short-term assessment results can be seen in Table **5.6**.

Table 5.6. Short-term health impact assessment results for nitrogen dioxide pollution.

RESULTS		Considered period	
		No. of cases per 10000 people	IC 95%
Scenario 1: Number of cases attributable compared to a low level of pollution	10 $\mu g/m^3$	10.72	7.48 - 13.97
Scenario 2: The health gain attributable to a suppression of pollution levels above	40 $\mu g/m^3$	1.46	1.02 - 1.90
Scenario 3: The health gain achieved for annual average exposure reduced by 25%	25%	4.12	2.89 - 5.35

The number of cases calculated with this tool should be interpreted as an estimate of the number of cases avoided per 10000 people under presented scenarios. This should allow better consideration of health effects in decision making and should contribute to set new air quality guidelines in line with public health objectives.

Adding GIS Capabilities

Health researchers and planners can benefit more from the capabilities of a geographic information system. GIS technology can be used for selecting study areas, identifying major polluters, assessing demographic distribution, creating dispersion and impact maps and for dissemination purposes by creating a wide variety of thematic maps. These facilities can all be integrated into the health risk assessment process.

As an example, to fulfil the hypothesis of homogeneous air quality for the selected area, GIS maps and satellite images can be used to assess the continuity of urbanization (Fig. **5.14**) and to identify major pollution sources (Fig. **5.15**).

Starting from here, all health risk assessment steps described in this chapter can be implemented within a GIS application. By adding GIS layers with demographic and pollution data and by using dispersion models based on meteorological data [24], we can assess with high precision the pollutant distribution and the number of people affected.

Also, to further increase the accuracy of the assessment, we can divide the study area into smaller areas based on the result obtained by using the dispersion model. Even the mathematical calculation modules required for assessment can now be integrated into GIS applications, and by connecting with real-time updated air pollution databases, assessment results can be made available very quickly in a format easily understood by a wide range of users.

A health risk assessment model for air pollution based on GIS technology is proposed by the European Environment Agency [25], which is graphically represented in (Fig. **5.16**).

Fig. (5.14). Area selection for health risk assessment.

Fig. (5.15). GIS representation of the principal sources of industrial pollution.

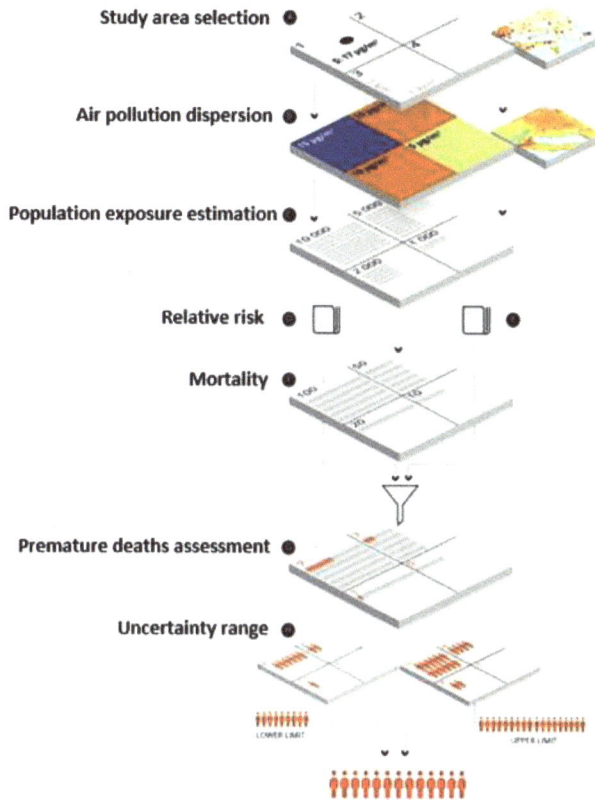

Fig. (5.16). EEA Health Risk Assessment Model using GIS technology [25].

CONCLUDING REMARKS

Due to the alarming increase of air pollution and the associated diseases, that can be observed in the last decades, the Assessment of the Health Impact and Health Risks of Air Pollution has become increasingly important. The presented theoretical information covers all the necessary Air Pollution Health Impact and Health Risks Assessment terminology and explains how to apply the assessment process and how to evaluate the risks at different scales. The need to better assess the air pollution related health risks has led to the development of a multitude of theoretical and numerical methodologies converted into a variety of health risks assessment tools using modern computer technology such as: automatic data acquisition processing and validation, relational database management systems, virtual instrumentation, and geographic information systems (GIS). The results provided by these tools can be transposed in local, regional, or national policies aiming to reduce the air pollution impact on human health.

REFERENCES

[1]   "World health organisation", Available at: https://www.who.int/health-topics/air-pollution#tab=tab_1

[2]   "United nations economic commission for europe", Available at: https://unece.org/air-pollution-a-d-health

[3]   *Joint EEA-JRC Report, Environment and human health* Publications Office of the European Union, No.5, 2013.

[4]   "World Health Organisation Health Impact Assessment", Available at: http://www.who.int/hia/en/

[5]   "World Health Organisation", Available at: https://www.who.int/tools/health-impact-assessments

[6]   ECHP, *Health Impact Assessment: Main concepts and suggested approach (Gothenburg Consensus Paper) (PDF).* European Centre for Health Policy: Brussels, 1999.

[7]   "United states environmental protection agency", Available at: https://www.epa.gov/healthresearch/health-impact-assessments

[8]   *WHO, Health risk assessment of air pollution : General principles* World Health Organization, Regional Office for Europe., 2016.

[9]   "United states environmental protection agency risk assessment for toxic air pollutants: A citizen's guide", Available at: https://www3.epa.gov/ttn/atw/3_90_024.html

[10]   "Government of Canada", Available at: https://www.ec.gc.ca/air/default.asp?lang=En&n=7E5E9F00-1#wsD88B8C88

[11]   World Health Organization, *World Health Organization* Air Quality Guidelines: Global Update, 2005.

[12]   "Concawe, An introduction to air qual", Available at: https://www.concawe.eu/wp-content/uploads/2017/09/DEF_AQ_AirQuality_digital.pdf

[13]   "United states environmental protection agency", *Dose-Response Assessment for Assessing Health Risks Associated With Exposure to Hazardous Air Pollutants.* Available at: https://www.epa.gov/fera/dose-response-assessment-assessing-health-risks-associated-exposure-hazardous-air-pollutants

[14]   E. Samoli, G. Touloumi, and A. Zanobetti, "Investigating the dose-response relation between air pollution and total mortality in the APHEA-2 multicity project", *Occup. Environ. Med.,* vol. 60, no. 12, pp. 977-982, 2003.
        [http://dx.doi.org/10.1136/oem.60.12.977]

[15]   *EU Air Quality Directives (2008/50/EC, 2004/107/EC).* Air quality guidelines: Global update, 2005.

[16]   "European commission air quality directives", Available at: https://environment.ec.europa.eu/topics/air/air-quality_en

[17]   "European environment agency air quality standards", Available at: https://www.eea.europa.eu/themes/air/air-quality-concentrations/air-quality-standards

[18]   "European environment agency air quality index", Available at: https://www.eea.europa.eu/themes/air/air-quality-index/index

[19]   T. Hassan Bhat, G. Jiawen, and H. Farzaneh, "Air pollution health risk assessment (AP-HRA), principles and applications", *Int. J. Environ. Res. Public Health,* vol. 18, no. 4, p. 1935, 2021.
        [http://dx.doi.org/10.3390/ijerph18041935]

[20]   "European environment agency", Available at: https://www.eea.europa.eu/data-and-maps/dashboards/air-pollutant-emissions-data-viewer-4

[21]   "European Environment Agency", *Health impacts of air pollution in Europe : Web report,* 2022.

[22]   "Health Monitoring Institute, APHEIS. Air pollution and health: A European information system", In: *Assessment of the health impact of air pollution in 26 European cities* Summary of European results and detailed results for French cities, 2002.

[23] Bontos M. D., Vasiliu D., Short-term Health Impact Assessment of Air Polution in Targoviste City (Dambovita County), Revista de Chimie, WOS: 000385266600040, ISSN 0034-7752, Vol. 67, no. 9, p. 1854-1859, 2016.

[24] D. Vasiliu, *Environmental monitoring.* Technical Ed, 2007.

[25] "European Environment Agency Assessing the risks to health from air pollution", Available at: https://www.eea.europa.eu/publications/assessing-the-risks-to-health

Environmental Risk Assessment, 2023, 111-139

CHAPTER 6

Risk-based Approach for Safe Drinking Water

Lăcrămioara D. Robescu[1,*]

[1] *Department of Hydraulics, Hydraulic Machinery and Environmental Engineering, Faculty of Energy Engineering, University POLITEHNICA of Bucharest, Bucharest, Romania*

Abstract: The United Nations (UN) recognizes access to safe drinking water as a fundamental human right. However, there are still many people without access to safe drinking water, and diseases caused by contaminated water pose a serious threat to human health. By 2030, progress would need to be made at four times the current rate to achieve UN Sustainable Development Goals related to water. The drinking water system has several points where undesired events could happen, allowing contaminated water to be delivered to the public. The safety and quality of the drinking water that they provide are always the suppliers' responsibility. Risk management is a crucial component in ensuring the supply of safe drinking water. One strategy for supplying consumers with safe drinking water is the "multiple barrier approach". Risk-based methodologies are more effective to identify and manage the hazards in the drinking water system to provide a consistent supply of safe drinking water. To encourage the development and use of risk management methods, the World Health Organization (WHO) has created guidelines for the quality of drinking water. European Drinking Water Directive that was revised and entered into force, which started in January 2021, also includes a risk management system for the "source-to-tap". This chapter presents aspects concerning water contamination and health, an overview of drinking water supply systems, safe drinking water risk management strategies, and the framework for safe drinking water to focus on water safety plan development.

Keywords: Safe drinking water, Water contamination, Drinking water supply system, Multibarrier approach, Risk-based approach, Water safety plan.

INTRODUCTION

Water is a precious resource that sustains life. It is used for various purposes such as domestic use, irrigation, industrial use, commercial use, power generation, aquacultural use or recreational use [1]. Some of these are more important than others. For example, having a minimum of 20 litres of water per person per day [2] is the most important need in contrast with all other needs, like recreational

---

* **Corresponding author Lăcrămioara D. Robescu:** Department of Hydraulics, Hydraulic Machinery and Environmental Engineering, Faculty of Energy Engineering, University POLITEHNICA of Bucharest, Bucharest, Romania; E-mail: diana.robescu@upb.ro

Diana Mariana Cocârță (Ed.)

use that is the least important in the hierarchy of water requirements [3]. We need water daily for many purposes, like drinking, food preparation, personal hygiene, washing, cleaning, watering plants, *etc.* "Water sustains life, but safe, clean drinking water defines civilization" [4].

There isn't a universally accepted definition of safe drinking water. Safe drinking water is a relative term, which depends on the standards and guidelines of a country. WHO defines safe drinking water as water that does not represent any significant risk to health over a lifetime of consumption [5]. Besides this, the presence of certain amounts of natural minerals and essential elements in safe water intended for human consumption is also important [6].

A continuous supply of safe drinking water is a basic human right that was recognized in 2010 by UN General Assembly resolution A/RES/64/292 [7], and UN 2030 Agenda for Sustainable Development and the associated Sustainable Development Goal 6 - Ensure availability and sustainable management of water and sanitation for all (SDG 6) formulated by United Nations in 2015 [8]. One of the eight targets of SDG 6 is focused on safe and affordable drinking water. Moreover, to accelerate the efforts towards meeting water-related challenges, the period of 2018-2028 was declared an International Decade for Action on "Water for Sustainable Development".

Freshwater represents only 2.5% of the total amount of water on Earth. Out of about 70% of Earth's freshwater, less than 1% is readily available for human use [10], and this is not evenly distributed throughout the world. Despite the efforts of the states that led to increasing proportion of the global population using safely managed drinking water, from 70% in 2015 to 74% in 2020, there are too many people who still lack safe water access, mainly in rural areas [11, 12], Fig. (**6.1**) By 2030, progress would need to be made at four times the current rate to achieve universal coverage [11].

Besides other pandemics during history, the COVID-19 pandemic highlights again the importance of the provision of safe water. Along with adequate sanitation and hygiene, it helps to prevent and protect human health during all infectious disease outbreaks.

WATER CONTAMINATION AND HEALTH

Water Contaminants of Health Significance

Diseases related to contamination of drinking water constitute a major threat to human health. In essence, there are four types of contaminants: inorganic contaminants, organic contaminants, biological contaminants, and radiological contaminants [13]. Contaminants can be derived from various sources in the urban water cycle. Some contaminants naturally occur in water, but others are the by-products of man-made processes or spread by human or animal wastes. They may cause adverse health effects from single exposures (*e.g.,* microbial pathogens) or long-term exposures (*e.g.,* many chemicals).

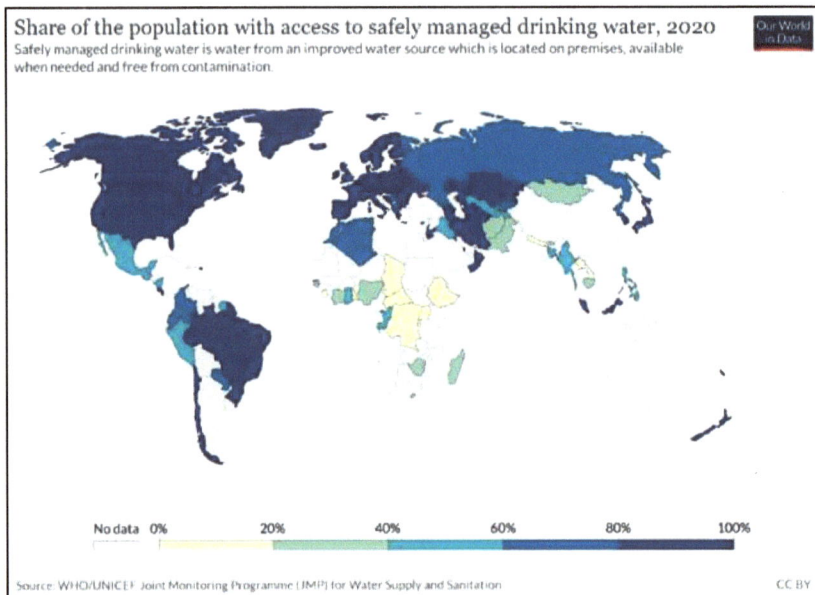

Fig. (6.1). Share of the population with access to safely managed drinking water [9].

Most of water-related health diseases are the result of microbial contamination (bacteriological, viral, protozoan, or other biological) or chemical contamination of drinking water. Microbial risks are associated with ingestion of water contaminated by faeces containing pathogens and by other microbial hazards (helminths, toxic cyanobacteria, and Legionella).

Chemical risks are related to chemicals that may occur in drinking water, such as fluoride, arsenic, uranium, selenium, nitrate, nitrite, and lead. They can cause adverse health effects after prolonged periods of exposure, less for short-term or single exposure [5].

Radionuclides can be naturally presented in water and thus, the risk associated with them should be also considered, although usually it is very small.

Depending on the water involvement in the transmission of diseases, there are four types of diseases ([1, 14]):

o Water-borne - infections spread through ingestion of contaminated water (*e.g.,* cholera, typhoid fever, viral hepatitis, viral gastroenteritis, giardiasis, ascariasis *etc.*)

o Water-washed - diseases due to the lack of proper sanitation and hygiene (*e.g.,* fungal skin diseases – ringworm, conjunctivitis, or infections – scabies, louse-borne epidemic typhus, COVID-19)

o Water-based - infections transmitted through parasites that spend part of their life cycle in water (*e.g.,* bilharzia, dracunculiasis)

o Insect-vector - diseases transmitted by insects that breed or feed in or near water bodies (*e.g.,* malaria, yellow fever); these diseases are not related to the lack of safe drinking water.

The presence and concentration of contaminants in water are tested using different microbial, chemical, or physical tests. However, it is important for safe drinking water supply to know the sources of contamination and how contaminants may enter the water supply.

Overview of the Drinking Water Supply System

Water utilities must continuously provide safe, accessible, acceptable, palatable, affordable, and reliable water in quantity required by consumers. Drinking-water suppliers are always responsible for the quality and safety of the water that they produce.

The drinking water supply system is a complex system of hydrologic and hydraulic components, which provides water supply for various consumers at a safe quality for human consumption, complying with legislative limits and aesthetically acceptable, that is water should be the view appeal and taste appeal for consumers.

The drinking water supply system starts with the abstraction of raw water from different sources (groundwater, streams, natural or artificial lakes, sea, and ocean), which is then transmitted to a treatment plant.

In the drinking water treatment plant, the raw water flows through a series of treatment processes to remove impurities and harmful substances from water. Conventional water treatment consists of several steps, Fig. (**6.2**): screening and straining, chemical addition, coagulation and flocculation, sedimentation, filtration, and disinfection. Typically, raw water enters a water treatment plant through an intake structure that may include transmission pipes, screens, strains, and pumping stations, depending on the water source and the size of the water supply [15]. **Screening and straining** are used to remove large and suspended debris from water (leaves, sticks, fish, trash, tree's trunks or limbs, algae, aquatic organism) that can impede or damage plant equipment. Furthermore, these items have no place in potable water [16].

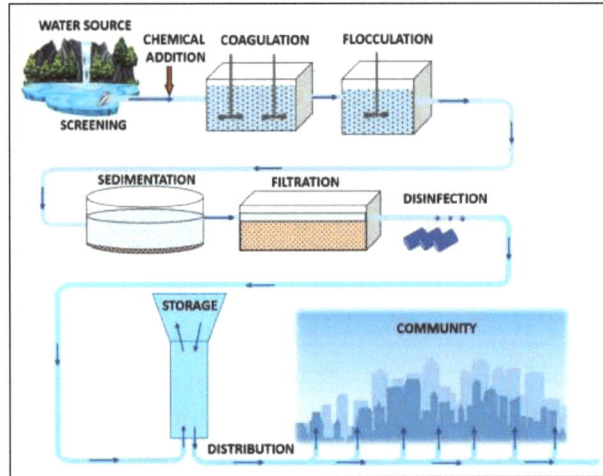

Fig. (6.2). Conventional drinking water treatment (Author's design of https://www.vecteezy.com/free-vector/water, https://www.vecteezy.com/free-vector/propeller, https://www.vecteezy.com/free-vector/building).

Coagulation - flocculation is applied to remove colloidal particles (light and fine non-sedimentable solid suspensions) or precipitable dissolved particles. These are a combination of biological organisms, bacteria, viruses, protozoa, zooplankton, colour-determining particles, and organic or inorganic substances, fine or dissolved precipitable.

The process includes two stages: coagulation is the rapid phase by which the colloidal particles are destabilized using a coagulant solution; flocculation is the slow phase through which the destabilized particles agglomerate and form larger flocs that can easily settle by adding flocculants or coagulation adjuvants. Commonly used coagulants are metal salts (aluminium sulphate, aluminium hydrochloride, ferric chloride, ferric sulphate, ferrous sulphate) and cationic, anionic, or non-ionic synthetic polyelectrolytes. Coagulants must be mixed very well with water to form a heavier floc. It can be used flash mixers for coagulation and paddles for flocculation that gently mix the water, or, instead of these, variable speed mixers.

Sedimentation aims to remove settleable particles by gravity. Once the flocculation is completed, the water containing flocs flows to the sedimentation tank. Water moves slowly through the sedimentation tank from the entrance point towards the discharge weir, allowing the flocs to settle. The sludge accumulated at the bottom of the tank is scraped by a rake, which continuously travels across the bottom of the tank in the case of circular tanks or along the bottom of the tank in the case of rectangular tanks. The sludge collected in a hopper is pumped to the sludge disposal facility.

Filtration is a physical process of separating suspended and colloidal particles from water by passing water through a granular material [16]. There are many types of filtration media: sand, anthracite, coal, or some other type of granular material. The most widely used are rapid or slow gravity sand filters. The water flows down from the top to the bottom of the filter, where it is collected in a drain system. Activated carbon filters may be used to remove organic compounds that can cause tasting or odour problems. Membrane filtration may be used in addition to or instead of conventional filtration. Filter membrane has very small pores, and it acts as a barrier which selectively separates the two phases by mass transfer, making it possible to separate the phases of a mixture. For water treatment, there are used pressure – driven processes: reverse osmosis (RO), nanofiltration (NF), ultrafiltration (UF), and microfiltration (MF). The membranes can be arranged in different configurations: flat-sheet, hollow-fibre, and spiral wound.

Disinfection is the selective destruction or inactivation of microbial pathogens, and it is essential in the supply of safe drinking water. This can be accomplished either physically (*e.g.,* using ultraviolet rays – UV) or chemically (*e.g.,* ozone, chlorine). The treated water is stored and finally distributed through the distribution system to the end users at the pressure, quality, and quantity required by them. Residual chlorine is used to assure that drinking water is safe at the consumer's tap. This remaining disinfectant kills the pathogens living as water flows through the distribution system. UV light and ozone disinfect water very well in the treatment plant, but they don't continue to kill pathogens in the distribution system. The use of chemical disinfectants usually results in the formation of chemical by-products, the so-called disinfection by-products (DBPs). In contrast to the problems associated with insufficient disinfection, the health concerns posed by these byproducts are, however, minimal [5].

The size and type of water treatment plant depend on several factors: the type of source or sources of water, the quality of the raw water, water quality standards, the water demand of the population served, available area for building the plant, the available area for waste disposal, fire protection, and investment and operational costs.

There are many parts of the drinking water system where undesired events may occur that could compromise the system, and contaminated water can be delivered to the public. Contaminants can get into drinking water at the water's source or in the distribution system after the water is treated in the treatment plant.

Numerous outbreaks are linked with distribution system faults, which can include cross-connections, back-siphonage, burst or leaking water mains, pressure fluctuations and leaching from pipework, contamination during storage, neglijent methods used to repair water mains and installation of new water mains, low water pressure and intermittent supply [17]. The successful operation and maintenance may assure the integrity of the drinking-water supply system and protection from contamination.

SAFE DRINKING WATER PROVISION AND RISK MANAGEMENT STRATEGIES

Urban water cycle risk management has been a rising concern for both water service companies and customers. Managing risk is an essential requirement to assure safe drinking water provision. Risk management is a process that includes identifying risks, analysing decision choices to manage risks, choosing amongst alternatives, acting to put the right measures in place, and monitoring and evaluating the results [18].

Multiple-Barrier Approach

One strategy for supplying consumers with safe drinking water is the "multiple barrier approach", in which different stages of the water supply system can serve as barriers. Every stage offers opportunities to manage water quality and several locations throughout the process—instead of only depending on a single supply system barrier—can prevent or decrease water quality risk, (Fig. **6.3**).

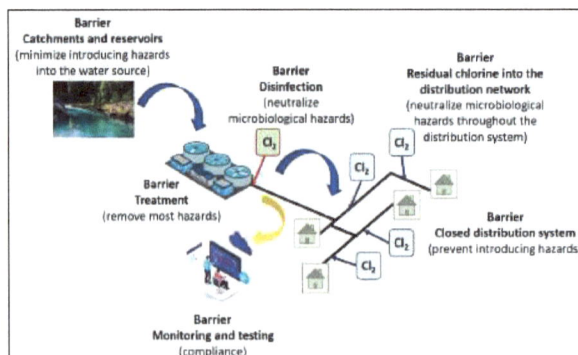

Fig. (6.3). Multiple-barrier approach (adapted from [19], using images from: Photo by Bob Wick, BLM, https://www.flickr.com/photos/blmoregon/19270292513; https://vecta.io/symbols/310/human-developmen--infrastructure/131/water-drinking-water-treatment-plant-1; https://www.vecteezy.com/free-vector/monito-ring).

The multi-barrier strategy includes three major parts: source water protection, drinking water treatment, and the drinking distribution system. In this approach, the so-called end-product testing, conducted right before or during distribution, ensures the quality of the water.

There are some drawbacks to this methodology [20]:

o The potential health problem can only be highlighted after the water has been consumed.

o Some water-borne pathogens cannot be detected or are insecurely detected using the classical indicator microorganisms for microbial contaminants.

o The number of samples for analysis is inadequate, given the overall amount of water supplied.

End-product testing must be carried out in conjunction with a management framework to ensure the safety of the final product. A thorough understanding of the risk of contamination as well as good risk analysis and management are necessary for the provision of safe drinking water.

Risk-Based Approach

Because diseases associated with drinking water contamination place a significant burden on human health, the World Health Organization (WHO) has developed standards for the quality of drinking water to promote the creation and application of risk management strategies that will assure the security of drinking water supplies through the control of hazardous water constituents.

A risk-based approach, which considers the characteristics of the drinking water supply from its catchment and source to its use by consumers, is proposed by the WHO [21]. This approach is named Water Safety Plan (WSP). An international group of experts created the Bonn Charter for Safe Drinking Water in 2004 as an add-on to the WHO standards to give a framework for managing water supplies from catchment to consumer. It provides the components to produce safe, high-quality drinking water that has consumers' trust when combined with WHO Drinking Water Quality Guidelines [22].

Risk-based methodologies to identify and manage the hazards in the drinking water system must be used to provide a consistent supply of safe drinking water. Additionally, they will boost customer confidence in the water they receive, which must encourage more people to drink tap water. This will benefit the environment by decreasing the use of plastics and greenhouse gas emissions [23].

For more than 30 years, the European Union has implemented policies to ensure that the population has access to safe drinking water. In December 2020, the European Parliament adopted the revised version of the European Drinking Water Directive that entered into force at the start of January 2021 with an implementation period of two years. In keeping with the European Green Deal's zero pollution objective for a toxic-free environment, it guarantees the greatest drinking water standards in the world.

The updated directive includes a few significant elements [6]:

o Stricter monitoring procedures should be applied to preserve drinking water quality by modifying treatment processes, with the goal of identifying any threats to drinking water early on.

o Watchlist mechanisms are being used to address the growing public concern over developing compounds' effects on human health when they are used in drinking water, such as pharmaceuticals, endocrine-disrupting chemicals, and microplastics, as well as to address new emerging compounds in the supply chain.

o Controlling all processes along the whole supply chain, from the catchment to the point of compliance, requires a comprehensive risk-based approach and a risk management system. The "source-to-tap" approach, which acknowledges the crucial part that the home distribution system plays in the water cycle, is what is known as this.

o Should be established testing and acceptance procedures for substances, compositions, and constituents to define specific minimum hygiene standards for materials intended to be used in the water cycle intended for human consumption.

o To reduce the lead content in water intended for human consumption, the components made of lead in domestic distribution systems can be replaced with materials that comply with the minimum requirements for materials that come into contact with water intended for human consumption.

o Treatment chemicals and filter media used in treatment and disinfection techniques for water intended for human consumption that are effective, safe, and properly controlled to avoid adverse effects on consumer health.

o A publicly accessible source of up-to-date, relevant information should be made available to the public in order to raise awareness of the implications of water use.

o To improve the efficiency of the water infrastructure and prevent overexploitation of water resources, water leaks should be assessed and fixed.

The risk-based strategy should consider risk assessment and risk management for the catchment areas, supply system and domestic distribution systems [6].

Tools for Risk Assessment of Drinking-Water Systems

To assess and address the risks associated with specific processes and to better understand, evaluate, characterize, communicate, and manage/control hazards, risk assessment and management are fundamental activities.

To perform risk assessment of drinking water systems, the tools available can be categorized into two main groups [24]:

o **Quantitative assessment**, that is based on models for generating a chain of events and estimating risk levels in numbers. This involves choosing assessment and measurement endpoints and comparing endpoint water quality measurements or distributions to a recommended value.

The value associated with the risk is a percentage that indicates the probability of the risk occurring or of it causing a specific negative effect on project objectives.

The models used for quantitative risk assessment can be [25]:

a. Point-estimate quantitative risk assessment model

Despite being particularly helpful in screening level assessments for individual hazards and endpoints, they do not adequately capture uncertainty and variability. (Quantitative microbial risk assessment/QMRA; Quantitative chemical risk assessment/QCRA);

b. Probabilistic quantitative model

They use randomised frequency distributions to represent one or more elements (Logic tree model: event tree, fault tree, Bayesian networks, and Markov model).

o **Qualitative assessment** that uses expert panels to rank issues with water quality based on their priority. These issues can be pollutants, pollution sources, or hazard events. (Examples: Hazard and Operability Study/HAZOP, Failure Mode and Effects Analysis/FMEA, Failure Mode, Effects and Criticality Analysis/FMECA).

The value associated with risk is the risk rating or scoring. A risk may be rated "Low" or given a score of 1 to indicate that the risk does not require immediate attention. Usually, a risk matrix is developed, which combines the consequences and the likelihood of a risk occurring.

The models used for qualitative risk assessment include [25]:

a. Conceptual descriptions of the cause-and-effect relationships that lead to risks arising from a particular activity or scenario.

b. Qualitative, subjective risk ranking models – they are used to rank scenarios, events, or options in terms of risk or impact rather than to provide estimates of actuals.

c. Semi-quantitative objective risk ranking models - are applied to ranking events, options, or scenarios, but they use objective data such as occurrence frequencies or receptor population size.

The field of risk analysis currently has a large variety of tools that have been used to either qualitatively or quantitatively compute and integrate the "probability of failure" and "consequences" at various levels in water supply systems [26].

While quantitative risk analysis is based on verified data, qualitative risk analysis is based on an individual's impressions or judgments. This is where qualitative risk analysis differs from quantitative risk analysis in that it tends to be more arbitrary. When there is a change in the perception of risk or when a new risk has been discovered, qualitative risk analysis should be carried out because it is generally simple, rapid, and inexpensive. When there is a lot of information about the risk and its consequences, as well as when the qualitative risk analysis needs to be validated, quantitative risk analysis should be conducted.

FRAMEWORK FOR SAFE DRINKING WATER

The management of risk is a major challenge for the planning, development, and operation of drinking water supply systems. Prior to implementing the most suitable risk management strategy, it must identify and assess the risks firstly, then provide strategies to minimize or mitigate the risks.

There are many parties involved in the provision of safe - drinking water, and each of them plays an important role, (Fig. **6.4**). They could influence or be impacted by the actions or decisions of the provider of drinking water. Therefore, a collaborative multiagency strategy to guarantee clean drinking water must be employed in preventive management [5].

Differentiating the functions and responsibilities of the service provider from those of the authority in charge of monitoring the quality of drinking water is an effective strategy for preserving public health. The term "dual-role approach" refers to the idea that monitoring and quality control of water should be handled by different, independent organizations.

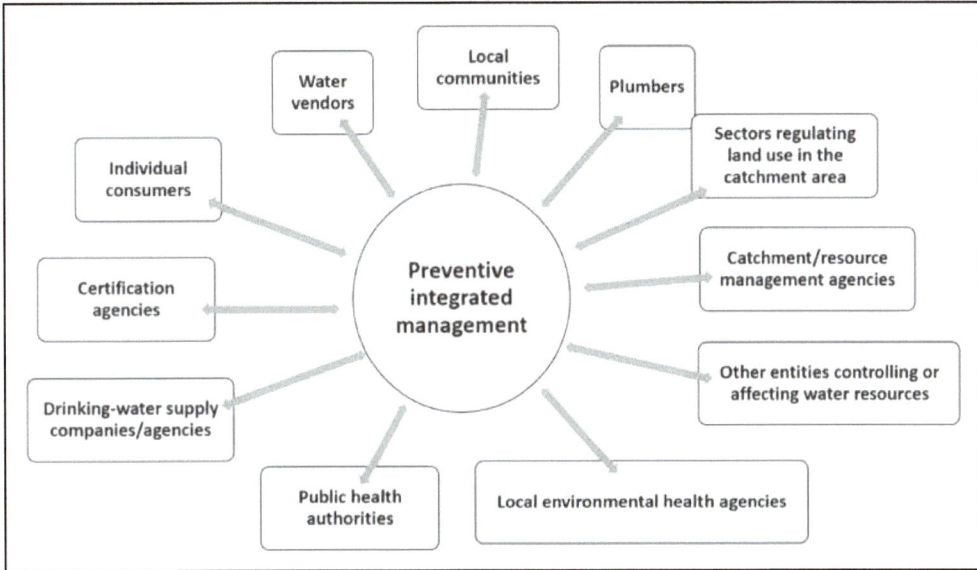

Fig. (6.4). Stakeholders involved in the provision of safe drinking water.

The development of a framework for safe drinking water is a fundamental and necessary condition to assure the safety of drinking water. Both the recommendations of the WHO framework for safe drinking water and European legislation are followed in this regard.

Key Components of the Framework for Safe-Drinking Water

The framework for safe drinking water was introduced by WHO in 2004, (Fig. **6.5**). It consists of five elements, and a water safety plan is created by combining three of them [5]:

Fig. (6.5). Framework for safe drinking water (adapted from [21]).

o *Health-based targets* are based on an evaluation of health concerns.

o *System assessment* ascertains if the entire drinking water supply from source to tap can produce water that fulfills the health-based targets.

o *Operational monitoring* of the control measures in the drinking water supply, which are crucial for ensuring the safety of the drinking water.

o *Management plans* (which detail the system evaluation and monitoring plans and specify the steps to be performed in both normal operation and emergency situations, including upgrading and improving the system), *documentation and communication.*

o *System of independent surveillance*, meaning a method of impartial monitoring that confirms the aforementioned are operating properly.

Health-based targets are established to safeguard and enhance public health. They should, therefore, be included in the broader public health policy and typically have a nationwide scope. Setting health-based targets requires high-level (national) health authorities to consider the general state of the public's health as well as diseases brought on by waterborne bacteria and pollutants. The health-based targets ought to be practical given the operational circumstances in the region. Different sorts of targets will be appropriate for varied objectives, therefore, in most countries, a variety of targets can be utilized. It may confer with other interested parties, such as water providers and impacted communities, while establishing these goals.

When setting targets, it must consider both water and other sources of microbial, chemical, or radiological hazards, including food, air, person-to-person contact, consumer products, inadequate sanitation, and poor personal hygiene. Targets must be realistic, measurable, based on scientific facts, and pertinent to local conditions (including economic, environmental, social, and cultural variables), as well as financial, technological, and institutional resources, to provide effective health protection and improvement [5].

There are four principal types of health-based targets ([21, 27]):

o *Health outcome targets* – they are set considering a quantifiable reduction of the disease, where adverse effects follow shortly after exposure (epidemiologically based) or they can be approximated based on information concerning exposure and dose-response relationships (risk assessment-based).

o *Water quality targets* – they are guideline concentrations of chemicals from water resources or from materials (chemical reagents or by-products) that pose a risk to human health when exposed over a long time and where concentration changes are small or persistent.

o *Performance targets* – they are set for constituents (microbial contaminants or chemical constituents), that pose a considerable risk to public health either by short-term exposure or significant variations in quantity or concentration over relatively short time periods. They are typically expressed in terms of of the amount of the problematic chemical that must be reduced, or the effectiveness of the method used to prevent contamination.

o *Specific technology targets* – they are targets set by national regulatory agencies for specific actions for smaller municipal, community and household drinking water supplies.

The burden of disease caused by various water-related risks needs to be quantified and compared, taking into consideration the differing probabilities, severity levels, and duration of impact. To allow for the implementation of a uniform strategy for each hazard, such a metric should be applicable regardless of the kind of hazard (microbial, chemical, or radioactive). In this regard, WHO uses the disability-adjusted life year, or DALY. DALYs can be used to define the tolerable burden of disease (an upper limit of the burden of health effects associated with waterborne disease that is established by national policymakers) or the reference level of risk (an equivalent terms used in the context of quantitative risk assessments).

Water Safety Plan, which is the responsibility of water suppliers, is a risk-based approach to ensure safe drinking water most effectively from source to tap, by preventing contamination of source waters, treating the water to meet the water quality targets, and preventing recontamination during storage, distribution, and handling of drinking water.

Three separate steps are used to achieve this, and they are directed by health-based targets and monitored by drinking water supply surveillance:

o *System assessment* – to establish whether the entire water supply system can deliver water with a quality that satisfies the health-based standards.

o *Operational monitoring* – to determine control methods and suitable means of operational monitoring to quickly detect any performance deviation and guarantee that the health-based targets are reached.

o *Management plans, documentation, and communication* – to outline what should be done under both normal operating settings and emergency situations, and documenting the system assessment, monitoring and communication plans and supporting programmes.

Surveillance is an ongoing and diligent public health assessment and overview of the safety and acceptability of drinking water supplies. It complements the water supplier's role in controlling water quality and spans the entire water supply system, from catchment to tap.

The a monitoring agency periodically checks the water quality to see if a waterborne disease outbreak is occurring and to take appropriate action. Additionally, it guarantees that any departure from the health-based targets is duly reviewed and remedied. Drinking-water supply surveillance does not take the place of or replace the duty of the provider of water to guarantee that the supply is safe to drink and complies with established health-based standards.

Development of Water Safety Plan

Water Safety Plan formulated by WHO in 2004 [21] shows how water supplier manages the public health risk that might arise from contaminated drinking water reaching the public. It is based on a multiple-barrier approach and Hazard Analysis and Critical Control Point (HACCP) ([27, 28]). WSP also uses elements of other approaches, such as those within ISO 9000 and the concept of total quality management (TQM).

WSP has three key components - system assessment, monitoring and management and communication - that can be divided into several steps, (Fig. **6.6**), [5, 27, 28].

• Assemble the Team

Creating a diverse, experienced team who have a complete understanding of the drinking water system concerned is the first step in building a WSP. Managers, engineers (operations, maintenance, design, capital investment), technical staff involved in water supply system operation, water quality controllers (microbiologists and chemists), representatives of relevant consumers, and representatives of relevant catchment-level agencies should all be included in it. Universities and other professional organizations should be included as independent members. The team leader should be carefully appointed because the role of the team leader is essential in the development and implementation of the WSP.

• System Assessment

To make sure that risks and hazards are properly evaluated and managed, an assessment of the type of water quality and the method utilized to produce the quality of water is crucial. It involves three steps:

o Document and describe the water supply system

o Hazard analysis

o Identify control measures

Document and Describe the Water Supply System

To give the team working on the WSP an overview of the supply and a basic grasp of the controls already in place, the first stage in the system assessment process is to completely describe the water supply from the source to the endpoint of supply, (Fig. **6.2**). The information offered ought to be adequate to pinpoint the system's weak points, relevant types of hazards, and control measures. Depending on how complex the system is, different water supply systems will require different levels of detail. Some suggestions for the main information are listed in Table **6.1**.

Table 6.1. Examples of water supply system information for the WSP team ([27, 29]).

Source	Catchment characteristics Raw water sources: surface, groundwater, roof catchments, reused wastewater Runoff and/or recharge processes Alternative water supplies
Treatment	Source abstraction Water transmission Water treatment (*e.g.,* process steps, treatment chemicals, microbiological log reduction, chemical reduction, *etc.*) Equipment Monitoring and automation
Distribution system	Regulatory water quality requirements Storage Pipe networks Construction materials Hydraulic conditions (pressure, flow, water age, *etc.*) Operation and monitoring Current/predicted climate influence on the distribution system
Consumer system	Users and uses of water separately for industrial and domestic users Identify vulnerable groups who have specific water quality requirements;

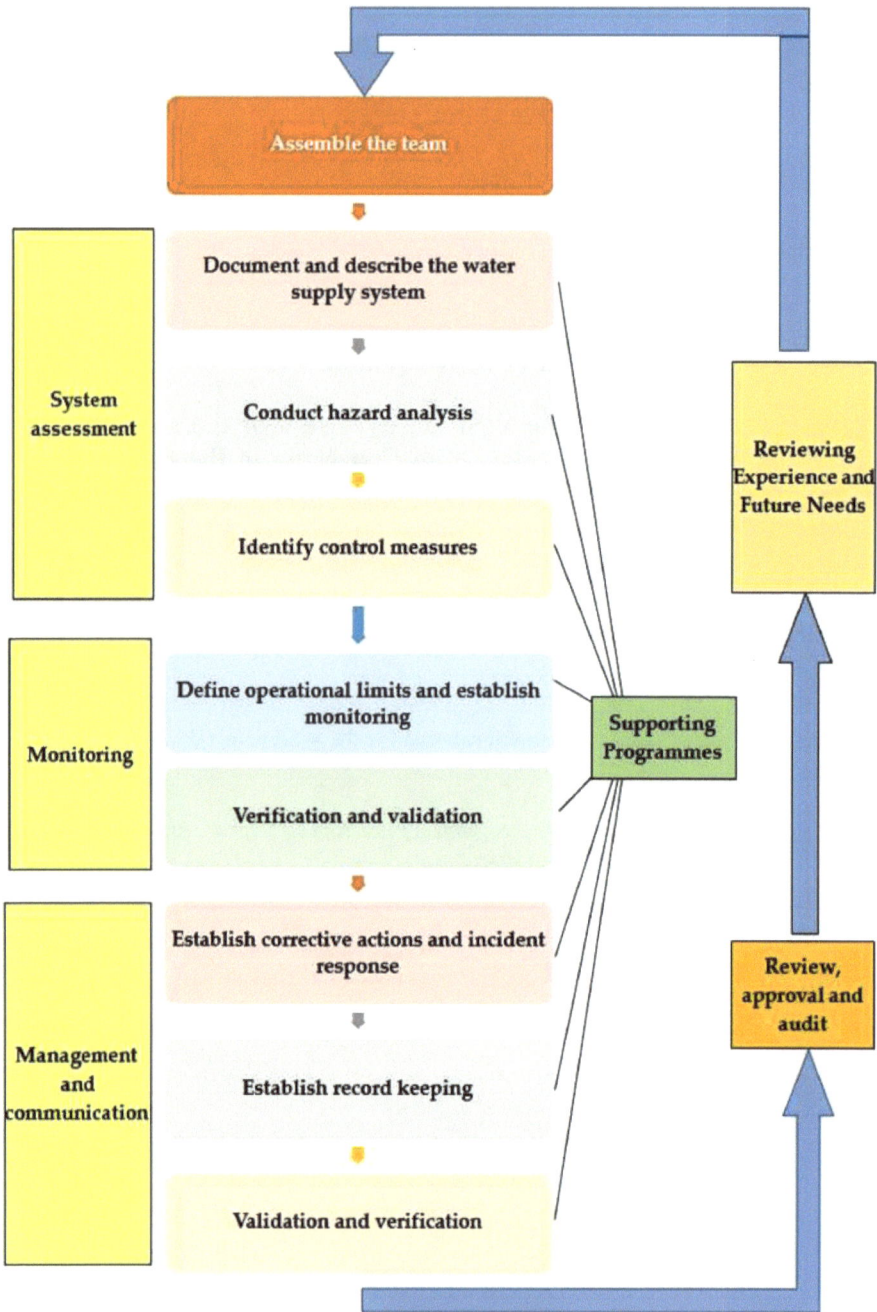

Fig. (6.6). Key components of WSP (adapted from [27]).

A current flow diagram for a drinking water supply and detailed information on catchments, surface water, groundwater systems, treatment systems, service reservoirs, and distribution systems should be included in this description. This will make it possible to identify risks, evaluate them, and manage them. An example is shown in Fig. (**6.7**).

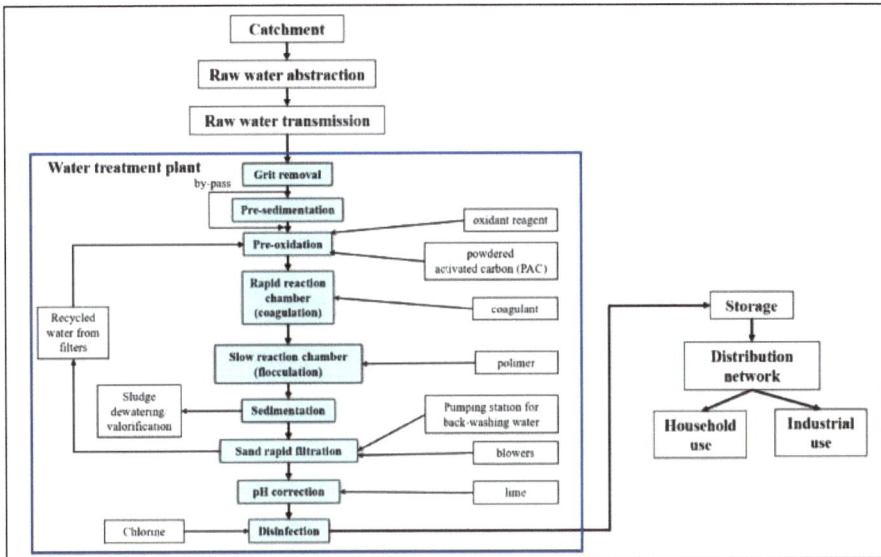

Fig. (6.7). Example of a flow diagram for a drinking supply system.

Hazard Analysis

Hazard analysis should identify potential hazards and hazardous events, assess risks, determine and validate existing control measures, and prioritise risks.

Hazards are defined as biological, physical, chemical, or radiological agents that can cause harm to public health [30].

Pathogens, such as bacteria, viruses, protozoa, and helminths, are examples of biological hazards. They can be enteric pathogens (*Escherichia coli (E. coli), Campylobacter, Cryptosporidium, Giardia*, and *Enterovirus*) or environmental pathogens (*Legionella, Non-tuberculous mycobacteria (NTM), Pseudomonas aeruginosa*). Non-pathogenic organisms that affect the acceptability of drinking water should also be considered.

Any chemical agent that could jeopardize the safety or appropriateness of water is termed a chemical hazard (*e.g.,* metals - manganese, aluminium, copper, lead, nickel, and cadmium or trihalomethanes- THMs).

Physical hazards can have an impact on water safety by creating a direct risk to health, decreasing the effectiveness of treatment, particularly residual disinfectants, or by causing users to utilize alternative, more contaminated water sources because they find the water unpalatable (*e.g.,* high water temperatures, high or low pH or high levels of turbidity).

Hazards may occur or be introduced throughout the water system, from the catchment to consumers.

Hazardous event is an event that introduces hazards to, or fails to remove them from, the water supply [30]. A place or circumstance that potentially results in a hazard is known as a ***hazard source***. Contamination can occur as a direct result of a hazardous event or indirectly. For clearly identifying hazardous events and associated hazards, it should be considered a cause-effect approach, using the basic formula, https://wsportal.org/what-are-water-safety-plans:

X(hazard) happens (on the water supply) because of **Y (hazardous event)**

Additionally, it should be clearly specified how the hazard is introduced [31]. For example, enteric pathogens got into the pipeline because of the burst of the old pipe of the water distribution, or harmful microbial contaminants remained in the water because of insufficient chlorine dosing due to dosing pump failure.

The evaluator should identify all hazards that could be associated with each step of the water supply or that could result in water contamination or interruption following the flow diagram previously prepared. Hazard identification should involve site visits, desktop review of system diagrams, and consideration of past events. Detailed information on identifying and managing risk for microbial and chemical hazards and hazardous events can be found in [32, 33]. Some examples of potential hazards affecting catchment, treatment, distribution, and consumer premises are presented in Table **6.2**, the more details are found in [30].

Table 6.2. Some examples of potential hazards within the water supply system.

Component of water supply affected	Hazardous event	Hazards
Catchment	Polluted runoff during rainfall	Contamination with enteric pathogens, pesticides, nitrates
	Inadequate or no sanitation	Microbial contamination with enteric pathogens
Treatment	Clogged filters	Organic matter and turbidity not removed
	Insufficient chlorine dosing	Microbial contaminants remaining in water
Distribution	Accidental pipe breakage	Contaminants getting into pipeline (microbial contaminants, pesticides, fuel oils etc.) Loss of supply
	Areas of low flow	Colonising of environmental pathogens and water contamination with them
Consumer premises	Lead plumbing, pipes, fittings and coatings	elevated lead concentrations in drinking-water
	Backflow due to unauthorized connections	Microbial and chemical contamination

Risk is defined as the probability that a hazardous event will occur within a given timeframe, along with how seriously it will affect the public's health.

Consequences may include injury to members of the exposed population as well as disruption of the water supply network. To differentiate between significant and less important hazards, the risk connected to each potential hazard event should be analysed, considering firstly the impact on public health.

The risk associated with each hazard may be described by the identified likelihood of occurrence (*e.g.,* 'certain', 'possible', 'rare') and evaluating the severity of consequences if the hazard occurred. (*e.g.,* 'insignificant', 'minor', 'moderate', 'major', 'catastrophic' *etc.*) [30]. It is very important to precisely define the meaning of the terms like 'rare', 'significant', 'major', *etc.* For example, 'Likely' means 'Occurs more often than once per month and up to once per week' or 'Moderate' means 'Possible extensive aesthetic problems or persistent exceeding of the maximum allowable value. Potentially harmful to a large population, but no mortality'. Detailed examples for defining these terms can be found in [29, 34].

To rank and prioritize the hazards, a risk assessment method should be decided: quantitative or semiquantitative, comprising estimation of likelihood/frequency and severity/consequence or a simplified qualitative approach based on the expert judgment of the WSP.

Risk is determined as the product of frequency and severity:

$$\text{Risk} = \text{Frequency} \times \text{Severity} \tag{6.1}$$

In the first ***initial risk assessment,*** there aren't considered any preventive control measures that are already in place.

Often a semiquantitative approach is chosen, and a risk assessment matrix is used, Table **6.3**.

An example of hazards' identification and associated risk assessment matrix is presented in Table **6.4**.

Identifying Control Measures

In the **second (residual) risk assessment**, for all identified hazardous events, *existing control measures should be identified in order to validate their effectiveness* to clearly determine the additional required control measures.

Table 6.3. Semi-quantitative risk approach matrix [30].

		Severity or consequences				
		Insignificant (no impact) Rating: 1	Minor (compliance impact) Rating: 2	Moderate (aesthetic impact) Rating: 3	Major (regulatory impact) Rating: 4	Catastrophic (public health impat) Rating: 5
Likelihood or frequency	Almost certain (once a day) rating: 5	5	10	15	20	25
	Likely (once a week) rating: 4	4	8	12	16	20
	Moderate (once a month) rating: 3	3	6	9	12	15
	Unlikely (once a year) rating: 2	2	4	6	8	10
	Rare (once every 5 years) rating: 1	1	2	3	4	5
Classification	Risk score	<5	5-9		10-15	>15
	Risk rating	Low (L)	Medium (M)		High (H)	Very high (VH)

Table 6.4. Example of the matrix for hazards' identification and risk assessment.

Process step	Hazardous event	Hazard	Likelihood	Severity	Score	Classification
Catchment	Polluted runoff with nitrates during rainfall due to improper use of fertilizers in the catchment area	Contamination with nitrates	3	4	12	H
Treatment	Insufficient chlorine dosing because of supply running out	Microbial contaminants remaining in water	2	5	10	H
Distribution	Old pipe leakage/burst	Contaminants getting into pipeline (microbial contaminants, pesticides, fuel oils etc.) Loss of supply	4	5	20	VH
Consumer premises	Backflow due to unauthorized connections	Microbial and chemical contamination	2	5	10	H

The multiple-barrier principle should be used for the identification and implementation of control measures. To evaluate the combined impact of various barriers, control measures must be identified both downstream and at the point of contamination, where the hazardous event occurs.

The first barriers in the protection of drinking water quality are provided by source and resource protection. Water treatment processes are the next barriers to drinking water system contamination. As the water is delivered to the consumer, the *distribution system* must offer a safe barrier against post-treatment contamination.

The effectiveness of each control measure is determined, that is, the validation of the control measure. For this purpose, it can be used as operating data, technical data from scientific literature, data from studies of pilot drinking water treatment plants or it can be needed as a monitoring program. Then, considering all currently in place control measures, the risks should be reassessed in terms of likelihood and effect Table **6.5**. Any risks that are still present after all control measures have been taken and are deemed unacceptable should be evaluated for potential additional corrective measures. Risks should be prioritized according to how they are likely to affect the system's ability to deliver safe water.

Table 6.5. Example of a matrix to identify and validate existing control measures, reassess, and prioritize risks.

Process step	Hazardous event	Hazard	Initial risks (any preventative control measures)				Existing control measures	Effectiveness of existing control measures				Residual risk (considering existing control measures)						Additional control measures
			Likelihood	Severity	Score	Classification		Yes	No	Somewhat	Validation of control measure	Likelihood	Severity	Score	Classification	Yes	No	Additional control mesures proposed
Catchment	Industrial wastewater discharged at upstream of intake	Chemical contamination	3	4	12	H	Water quality testing	-	-	x	Evidence of chemical contaminants levels	3	4	12	H	x	-	Water quality at the source should be routinely monitored to keep track of any pollutants contaminating the water source
Treatment	Insufficient chlorine dosing because of supply running out	Microbial contaminants remaining in water	2	5	10	H	Electronic stock control system with alarm on chlorine supply	-	-	x	Records on chlorine residual and chlorine stock	1	5	5	L	-	x	-
Distribution	Old pipe leakage/burst	Contaminants getting into pipeline (microbial contaminants, pesticides, fuel oils etc.) Loss of supply	4	5	20	VH	Rehabilitation or replacement of old pipes	-	-	x	Evidence showing old pipes replacements in case of breakdown	3	5	15	H	x	-	Evaluate the age and status of the pipes; Establish a program for continuous replacing the old pipes
Consumer premises	Backflow due to unauthorized connections	Microbial and chemical contamination	2	5	10	H	-	-	-	-	n.a.	2	5	10	H	x	-	Installation of backflow prevention devices; Develop and implement a policy for identifying and dealing with illegal connections; Implementing consumer education program

The drinking water supplier must indicate which level of risk it considers:

o Acceptable, for which it would not take any additional control measures to further reduce the risk.

o Unacceptable, for which it must take additional control measures.

o Unacceptable until the uncertainty is reduced, and for which it must take at least short-term control measures to reduce the risk.

The risk assessment is done based on the knowledge and data available or estimated. Consequently, uncertainties and limitations should be considered to determine responses to unacceptable risks.

Based on the risks identified in the previous step, a prioritized improvement/upgrade plan should be drawn up for each significant uncontrolled risk, considering short-, medium- or long-term activities. The risks should be recalculated and reassessed if new control measures introduce new risks. Monitoring the implementation of this plan is necessary to ensure that the changes that have been made are working, and that the WSP has been updated appropriately.

• Operational Monitoring

Operational monitoring evaluates at appropriate time intervals the effectiveness of control measures to enable effective system management and guarantee the accomplishment of health-based goals. In this regard, three steps should be followed: define operational limits, establish monitoring, and verifying the effectiveness of the WSP.

Theoretical and/or empirical studies can be used to establish a connection between the effectiveness of hazard control and the performance of the control measure, based on the parameters selected for operational monitoring. These parameters can be operational parameters (*e.g.,* pH, chlorine residuals, dissolved oxygen, hydraulic pressure, *etc.*) or observable parameters (*e.g.,* integrity of vermin proof netting, integrity of the fence around the wellhead, *etc.*).

To determine whether a control measure is working as designed, it is first necessary for each control measure to specify its ***operational limits****(or critical limits)*. The parameters used as limits should be directly or indirectly readily measurably, provide timely performance feedback, and allow for an appropriate response [27]. Limits can be upper limits, lower limits, a range, or an envelope of performance measures. They can be alert or action limits.

When the operational limit is exceeded, it signifies that something needs to be done to keep the control measure from deviating from compliance. Establishing the "what," "how," "when," and "who" principles is essential to monitoring Table **6.6**. It is important to create a monitoring plan and keep track of all monitoring activities.

Table 6.6. Example of operational monitoring and corrective actions.

Process step/ Hazardous event	Control measure	What?	Where?	When?	How?	Who?	Critical limit	Corrective action
Catchment *Industrial wastewater discharged at upstream of intake*	Water quality at the source should be routinely monitored to keep track of any pollutants contaminating the water source	Regularly monitor water quality at the source	Sampling point near intake	Hourly	Laboratory testing	Laboratory technician	e.g. acceptable value for microbial parameter/acceptable value for chemical parameter etc.	Alternative source using until the level is within acceptable limit Implementation of online monitoring with alarm triggered for unacceptable levels of contaminants

The effectiveness of the WSP is established through ***validation and verification***. These will confirm, by providing evidence, that the WSP is appropriate, implemented, and working properly, and water quality meets defined targets.

Validation ensures that the scientific and technical information supporting the WSP is correct. Various sources, such as the scientific literature, regulation and law departments, historical data, professional bodies, or supplier knowledge, can provide evidence for the validity of the water safety plans [27].

Three activities should be undertaken to accomplish WSP verification monitoring:

o *Compliance monitoring* is undertaken based on a monitoring plan. Each control measure must be monitored at a specific time interval to validate the performance against set limits. In case of unexpected results, corrective actions must be put in place according to a corrective action plan developed by the WSP team.

o *Consumer's satisfaction* is a valuable tool for detecting some problems related to water quality A procedure should be established to handle consumer's complaints (the modality to make complaints – *e.g.*, call center, email, social media, register complaints and identification data of complainants, review complaints, staff responsible, *etc.*).

o *Audits* should be undertaken on a regular basis to review operational records involving internal and external reviews. WSP auditing guidance can be found in [35].

• **Management and Communication**

Management Procedures

Effective management entails setting management procedures that define the steps to be taken in response to variations that occur under *standard operating conditions (Standard Operation Procedures)*, the steps to be taken in *specific "incident" situations,* and the steps to be taken in *unexpected and emergency situations* [30].

A *corrective action plan* should be developed to define the actions to be taken when the results of monitoring indicate a deviation from an operational limit.

Moreover, it is advisable to have a *general emergency response plan* in place in case of incidents and emergencies.

Supporting Programmes

Supporting programs are crucial for guaranteeing the safety of drinking water but do not directly impact the drinking water quality. Some examples of supporting programmes are training and continued education courses for personnel awareness and educational programs for communities, research and development, preventive maintenance, hygiene and sanitation, *etc.*

Management of Documentation and Records

Effective drinking water quality management systems are established and maintained based on appropriate documentation. As a result, record-keeping and documentation of processes and procedures are necessary for all WSP components. Documentation of a WSP should include [30]:

o Description and assessment of the drinking-water system.

o The plan for operational monitoring and verification monitoring of the drinking-water system.

o Water safety management procedures for normal operation, corrective action plan and incidents and emergency situations, including communication plans.

o Description of supporting programmes.

Documents must be periodically examined and amended as necessary. A document control system is needed for effective document management to make sure that the most recent versions are used and that older versions are deleted.

Records are necessary to evaluate the WSP's effectiveness and show that the drinking water system can implement and complies with the WSP. They must be properly stored to prevent damage and frequently back-up.

Generally, the records kept are [30]:

o Supporting documentation for developing the WSP.

o Records and results generated through operational monitoring and verification monitoring.

o Outcomes of unforeseen situations investigations.

o Documentation of management procedures.

o Records of training and continued education programmes for personnel.

Communication Strategy: Development is important in both WSP management procedures and surveillance implementation, considering inside and outside target groups. It should include:

o Communication plan required for the management of incidents and emergencies, which includes procedures for rapidly alerting the public health authorities and others to any important incidents involving the drinking water supply.

o Communication plan for the consumer and media that contains a summary of information to be made available to them at an appropriate time after the incident, but also meaningful information for surveillance purposes (*e.g.,* annual reports published on the supplier webpage, daily data regarding water quality, *etc.*).

o Procedures to receive and promptly handle community complaints.

• **Water Safety Plan Review, Approval and Audit**

WSP is reviewed by an appropriate body that should come to one of the following conclusions:

o WSP is approved in full and is ready for implementation.

o WSP receives provisional approval and can be implemented subject to ensuring the identified information gaps are filled.

o WSP is rejected as inadequate, and the supplier is required to go back and develop a new WSP.

A WSP should be periodically reviewed and updated, considering the results of monitoring processes. Moreover, this should be done in case of significant changes within the water supply system (for example, after an incident). The audit is undertaken internally by WSP teams and externally by regulatory authorities or qualified independent auditors.

CONCLUDING REMARKS

Access to safe drinking water is one of a basic human right, being of utmost importance for human health, disease prevention, hygiene, economic development, and environmental sustainability. There are many points in the water supply chain where drinking water can be contaminated after it is treated. Drinking water suppliers are responsible with the safety and quality of water that they provide. This chapter covers the main steps applied in the development of a risk-based approach for safe drinking water, based on the framework of safe drinking water recommended by WHO. Examples are provided for analyzing the various hazards and their associated risks from the catchment and source to the consumers, and for implementing appropriate measures to control and mitigate those risks. Implementing a risk-based approach for safe drinking water can guarantee that the population has access to safe drinking water of the quality in line with standards and guidelines of each country.

REFERENCES

[1]   W.A.S.H. Open, "Open wash, urban water supply, the open university UK/world vision Ethiopia/UNICEF", Available at: https://www.open.edu/openlearncreate/course/view.php?id=2099.

[2]   "United Nations Development Programme (UNDP), human development report 2006: beyond scarcity: power, poverty and the global water crisis,", Available at: http://hdr.undp.org/ en/ content/(Accessed on: 2022).

[3]   "WHO/WEDC technical notes on wash for emergencies, technical note no.9", Available at: https://cdn.who.int/media/docs/default-source/wash-documents/who-tn-09-how-much-water-is-need-ed.pdf?sfvrsn=1e876b2a_6

[4]   J.D. Brookes, and C.C. Carey, "Goal 6—rising to the challenge: Enabling access to clean and safe water globally, UN Chronicle, No. 4, Vol. LI, 20135", Available at: https://www.un.org/en/chronicle/ article/goal-6-rising-challenge-enabling-access-clean-and-safe-water-globally#:~:text=Water%20sus-tains%20life%2C%20but%20clean,of%20the% 20world's%20poorest%20nations

[5]   "World Health Organization (WHO), Guidelines for drinking-water quality: Fourth edition incorporating the first addendum, Geneva. License: CC BY-NC-SA 3.0 IGO", Available at: https://www.who.int/publications/i/item/9789241549950.

[6]   "DIRECTIVE (EU) 2020/2184 of the european parliament and of the council on the quality of water intended for human consumption (recast), official journal of european union, L453/1-L435/62", Available at: https://eur-lex.europa.eu/legal-content/EN/TXT/PDF/?uri=CELEX:32020L2184&from =EN

[7]   "United Nations (UN), resolution adopted by the general assembly on 28 july 2010. 64/292. the human right to water and sanitation, sixty-fourth session, agenda item 48, A/RES/64/292", Available at: https://www.un.org/en/ga/64/resolutions.shtml.

[8]   "United Nations (UN), Resolution adopted by the General Assembly on 25 September 2015. 70/1. Transforming our world: the 2030 Agenda for Sustainable Development, seventinth session, Agenda items 15 and 116", Available at: https://sdgs.un.org/2030agenda Available at: https://www.un.org/en/ga/64/resolutions.shtml

[9]   H. Ritchie, and M. Roser, "Clean water and sanitation", Available at: https://ourworldindata.org/clean-water-sanitation (Accessed on: Mar. 4 2022).

[10]   U.S. Geological Survey (USGS)., "Water science school, how much water is there on earth?", Available at: https://www.usgs.gov/special-topics/water-science-school/science/how-much-water-there-earth

[11]   United Nations (UN), "The Sustainable Development Goals Report", Available at: https://unstats.un.org/sdgs/report/2022/)(Accesed on:17.08 2022).

[12]   "World Health Organization (WHO) and the United Nations Children's Fund (UNICEF), Progress on household drinking water, sanitation and hygiene 2000-2020: Five years into the SDGs, Geneva, Licence: CC BY-NC-SA 3.0 IGO", Available at: https://www.unwater.org/app/uploads/2021/07/jmp-2021-wash-households-LAUNCH-VERSION.pdf

[13]   S. Sharma, A. Bhattacharya, "Drinking water contamination and treatment techniques", Applied Water Science, vol. 7, 1043–1067, 2017. [Online]. Available: https://ourworldindata.org/ clean-wate-sanitation [Accessed Sep. 5, 2022].

[14]   "Water engineering and development center (WEDC), Water : Quality or quantity?, mobile notes 60, WEDC, Loughborough University, 2017", Available at: https://wedc-knowledge.lboro.ac.uk/resources/e/mn/060-Water-quality-or-quantity.pdf

[15]   S.N. Foellmi, "Intake Structures", In: *Water Treatment Plant Design.*, E.E. Baruth, Ed., 4th. McGraw-Hill: New York. Available at: https://www.accessengineeringlibrary.com/binary/mheaeworks/9cae858b082ab522/001db5980b4444ab0c93668b680698e8283d13700f25427131ec4ef6f8a88745/intake-structures.pdf?implicit-login=true&sigma-token=uSmkhhuU0cgUcjmxAeTmAI5wnIFj1wo2sW H6MSaw8oU(Accessed on:Sep.4 2022).

[16]   F.R. Spellman, *Handbook of Water & Wastewater Treatment Plant Operations* Lewis Publishers, CRC Press: Boca Raton, 2003, p. 696. [http://dx.doi.org/10.1201/9780203489833]

[17]   "World Health Organization (WHO) Water safety in distribution system", Available at: ps.who.int/iris/handle/10665/204422 (Accessed on:Feb. 20 2022).

[18]   A. Bendz, and Å. Boholm, "Drinking water risk management: Local government collaboration in West Sweden", *J. Risk. Res.*, vol. 22, pp. 674-691, 2019.6 Available at: https://www.tandfonline.com/doi/full/10.1080/13669877.2018.1485168 (Accessed on: 2022). [http://dx.doi.org/10.1080/13669877.2018.1485168]

[19]   J.M.P. Vieira, "Water safety plans: Methodologies for risk assessment and risk management in drinking-water systems", *The fourth inter-Celtic Colloquium on Hydrology and Management of Water Resources* Guimaräes, Portugal 2005 Available at: https://www.aprh.pt/celtico/PAPERS/RT2P3.PDF (Accessed on:Sep. 6, 2022)

[20]   W.Q.M. Hunterwater, "A multiple barrier approach factsheet", Available at: https://www.hunterwater.com.au/documents/assets/src/uploads/documents/Fact-Sheets/Water-Quality/water_quality_mng_

multi_barrier_approach.pdf (Accessed on:Feb. 20, 2022)

[21] World Health Organization, "Water, sanitation and health team, guidelines for drinking-water quality", Available at: https://apps.who.int/iris/handle/10665/42852 (Accessed on:Feb. 20, 2022).

[22] "International water organization (IWA)", Available at: https://iwa-network.org/wp-content/uploads/2016/06/Bonn-Charter-for-Safe-Drinking-Water.pdf (Accessed on:Jan. 15 2022)

[23] A. Davison, G. Howard, M. Stevens, P. Callan, L. Fewtrell, D. Deere, and J. Bartram, *Water Safety Plans.*. Available at: https://apps.who.int/iris/handle/10665/42890 (Accessed on:Jan. 15 2022).

[24] "Safer drinking water for all europeans", Available at: https://ec.europa.eu/commission/presscorner/detail/en/IP_18_429 (Accessed on:Dec. 3 2022).

[25] A.S. Niedbalski, and V.V. Cos, *Risk assessment in drinking water supplies of Sweden and Latvia. An overview within the Water Safety Plan framework* chalmers university of technology: Gothenburg, Sweden , 2015. Available at: https://publications.lib.chalmers.se/records/fulltext/224342/224342.pdf

[26] R. Miller, J. Guice, and D. Deere, *Risk Assessment for Drinking Water sources. Research Report 78.* CRC for Water Quality Research Australia: Adelaide, 2009. Available at: https://web.archive.org/web/20190326230335/https://www.waterra.com.au/_r6852/media/system/attrib/file/1614/RR%2078%20Risk%20Assess%20DW%20sources.pdf (Accessed on:Feb. 20 2019).

[27] G.O.M.K. Mpindou, I.E. Bueno, and E.C. Ramón, "Risk analysis methods of water supply systems: comprehensive review from source to tap", In: *Appl. Water. Sci.* vol. 12. Springer, 2022, p. 56. [http://dx.doi.org/10.1007/s13201-022-01586-7]

[28] "World health organization (who) and international water association", Available at: https://www.who.int/publications/m/item/water-safety-plans-training-package (Accessed on:May 3 2022).

[29] "World Health Organization & International Water Association", *Water safety plan manual: step-by step risk management for drinking-water suppliers.* Available at: https://apps.who.int/iris/handle/10665/75141 (Accessed on:May 3 2022).

[30] "Ministry of Health Handbook for Preparing a Water Safety Plan", Available at: https://www.taumataarowai.govt.nz/assets/Uploads/Guidance/Handbook-Preparing-water-safety-plan-May-2019.docx (Accessed on:Oct.21 2022).

[31] S. Godfrey, and G. Howard, *Water Safety Plans (WSP) for Urban Piped Water Supplies in Developing Countries, WEDC.* Loughborough University: UK, 2004. Available at: https://sswm.info/sites/default/files/reference_attachments/GODFREY%20and%20HOWARD%202004%20WSP%20Developing%20Countries.pdf (Accessed on:May 3 2022).

[32] "Water Safety Plan", Available at: http://binmaleywaterdistrict.gov.ph/wp-content/uploads/2019/01/Water-Safety-Plan.pdf (Accessed Nov. 18 2022).

[33] A. Dufour, M. Snozzi, W. Koster, J. Bartram, E. Ronchi, and L. Fewtrel, "Assessing microbial safety of drinking water: Improving approaches and methods", Available at: https://www.who.int/publications/i/item/9241546301 (Accessed on:Sep. 7 2022).

[34] T. Thompson, J. Fawell, S. Kunikane, D. Jackson, S. Appleyard, P. Callan, J. Bartram, and P. Kingston, "Chemical safety of drinking water: assessing priorities for risk management", Available at: https://apps.who.int/iris/bitstream/10665/43285/1/9789241546768_eng.pdf(Accessed on:Nov.18 2022).

[35] "World health organization (who) and international water association", Available at: https://www.who.int/publications/i/item/9789241509527 (Accessed on:Nov.18 2022).

CHAPTER 7

Application of Risk Analysis to Waste Technologies

C. Stan[1,*]

[1] *Department of Energy Production and Use, Faculty of Energy Engineering, University POLITEHNICA of Bucharest, Bucharest, Romania*

Abstract: The current chapter presents the main municipal solid waste management processes used worldwide with potentially negative effects on human health: incineration, landfilling and composting by analyzing the defining elements regarding the main pollutants generated by the waste management processes in the form of solid, liquid, and gaseous discharges. Waste production, management and disposal involve more complex activities, with different potentials to affect health directly and indirectly through many pathways and mechanisms. The impact of waste may vary depending on numerous factors, such as the type of waste management processes, characteristics, and habits of the exposed population, duration of exposure, prevention, and mitigation interventions. Improper waste management in terms of health impact could be directly linked, to potential adverse substances, which leads to increased risk of cancer and quality of life decreasing or indirectly, to the environmental impact of the process, such as the contribution to global warming, loss of biodiversity and the depletion of non-renewable resources.

Keywords: Municipal solid waste, Waste management processes, Risk analysis, Risk assessment, Waste technologies, Landfilling, Thermal treatment, Composting.

INTRODUCTION

The three main methods for managing municipal solid waste that may have a negative impact on human health—incineration, landfilling, and composting—are discussed in this chapter. The provided information defines the key pollutants produced by waste management processes in the forms of solid, liquid, and gaseous discharges.

The management of waste has significant implications for environmental preservation and human health and welfare, sustainability, and economy.

---

[*] **Corresponding author C. Stan:** Department of Energy Production and Use, Faculty of Energy Engineering, University POLITEHNICA of Bucharest, Bucharest, Romania; E-mail: stan.constantin@yahoo.com

The inappropriate management of municipal solid waste presents some concerns for soil, water, and air pollution, which ultimately impacts the health of the people, as the World Health Organization (WHO) has frequently emphasized.

Municipal Solid Waste (MSW) generation is predicted to increase by up to 3.4 billion tons by 2050 [1], and even if waste management practices tend to improve, this happens from high-income to low-income countries. Due to the continued use of the most potentially hazardous waste management practices, such as open dumping and waste burning, the associated health risks are therefore, higher in low-income nations.

The impact of waste may differ based on a few variables, including the type of waste management methods, the characteristics and behavior of the population exposed, the length of exposure, and measures for prevention and mitigation.

A risk analysis of municipal solid waste begins with hazard identification and exposure assessment [2, 3]. The relationships between municipal solid waste management processes and potential adverse health effects are shown schematically in Fig. (**7.1**). The risks associated with waste management are visible, as are the potential environmental pathways through which the most exposed or vulnerable populations may ingest toxins.

Different waste management techniques lead to the generation of different compounds, as well as to different environmental impacts through exposure and transport. For example, in the direct incineration of waste, air is the first transport route to the environment. Dioxins, benzene pesticides, PCB_s, and other organic chemicals may be generated, and consumption of contaminated food may be an indirect source of exposure [4]. Pollution of groundwater by leaking leachate from the disposal of waste in landfills or open dumps may also affect drinking water [5, 6]. In this case, ingestion of water contaminated with harmful or carcinogenic substances would represent the subsequent exposure [7].

Some suspected effects of landfills and incinerators have been noted in research, such as a higher incidence of cancer and congenital anomalies and malformations in surrounding communities [8].

For the composting processes, in terms of health outcomes, some bioaerosols exposure can be found but with the mention to support a precautionary approach with no increased risks [9].

The Waste Framework Directive - Directive 2008/98/ EC outlines the basic information about the principles of waste management: waste must be managed without posing a risk to human health or the environment, without posing a risk to water, air, soil, plants, or animals, without causing problems due to noise or odor, and without affecting the landscape or places in the surrounding area [10]. The waste management hierarchy, arranged from top to bottom in Fig. (**7.2**), forms the basis for waste legislation and policy.

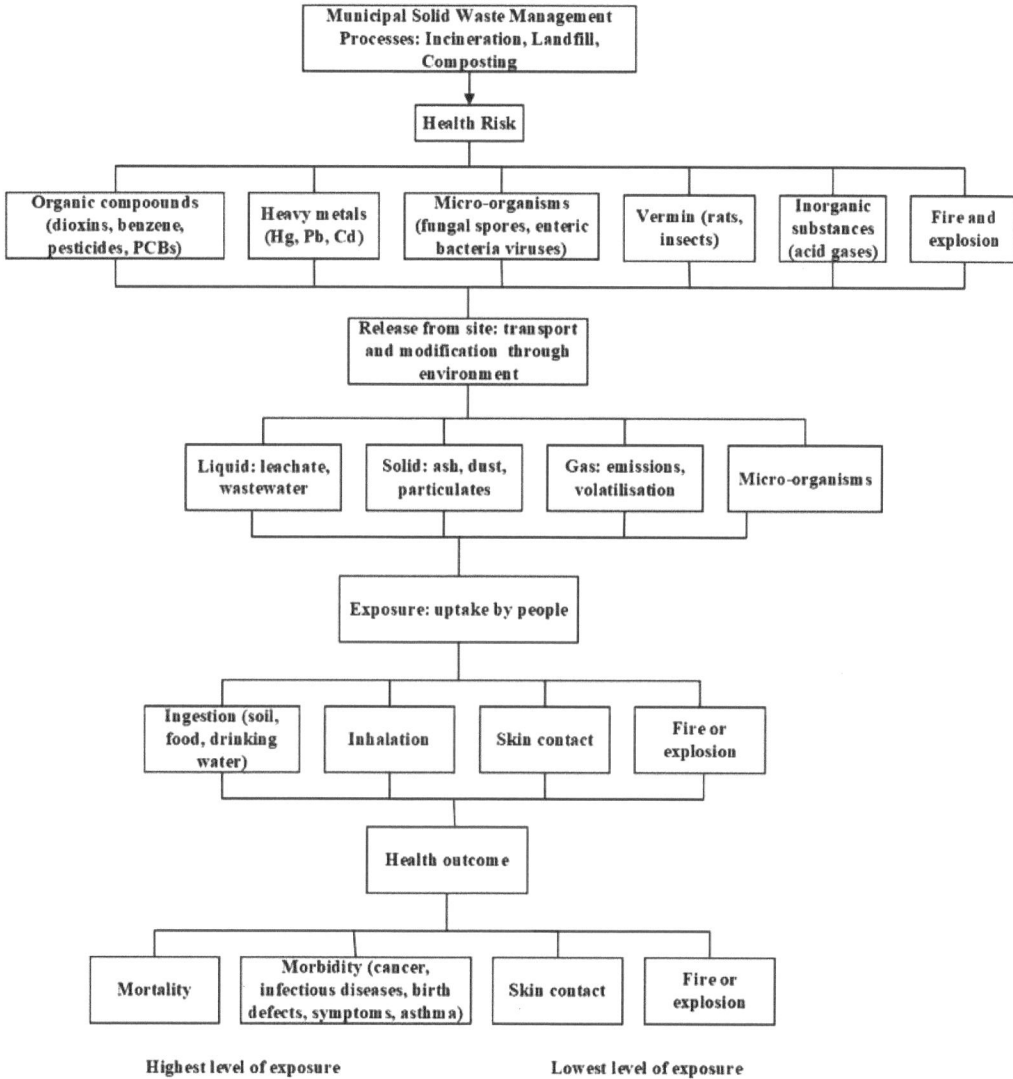

Fig. (7.1). Pathways from health hazards to health impacts of the municipal solid waste management processes.

According to recent studies, inadequate waste management techniques have been associated with a few contamination incidents, raising public concern about the lack of regulations, inadequate legislation, and the impact on the environment and human health [11, 12]. Therefore, a hierarchy of waste management practices based on the least environmental impact gives preference to waste reduction and prevention, waste reuse, recycling, and recovery.

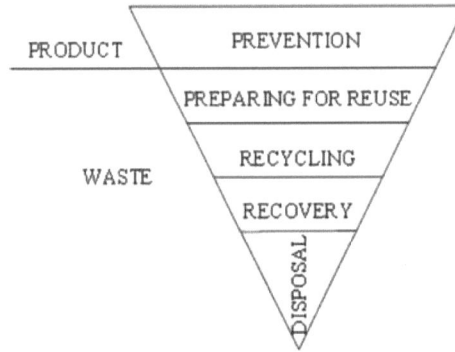

Fig. (7.2). EU hierarchy for waste management.

HUMAN HEALTH IMPACT OF WASTE MANAGEMENT PROCESSES

Health Effects of Waste Management

To determine the public health impacts, waste management strategies adopted locally, regionally, and nationally are needed. The potential negative effects of various substances that increase cancer risk and reduce the quality of life could be directly associated with improper waste management, in terms of health effects, or indirectly through the environmental impact of the process, such as its contribution to global warming, loss of biodiversity, and depletion of nonrenewable resources.

Untreated municipal waste can harm the health of people, especially those who live near the landfill. Leaks from the waste can contaminate soil and water streams and release heavy metals and persistent organic pollutants (POPs) that can cause air pollution [13, 14]. An effective waste management plan is critical for health reasons. Despite increasing recycling efforts, landfills and incinerators remain the primary methods of waste disposal worldwide [15]. Each step of waste handling, treatment, and disposal must be considered to assess the health risks associated with waste management practices.

Population Exposed to Pollution from Waste Management

When discussing the human health impact of waste management processes, a relevant factor is how much, and which population is exposed to these risks. Only a tiny portion of the population living near waste disposal facilities is exposed to pollution from these facilities, and not all residents of an urban area. Numerous studies in Europe, such as SESPIR and INTARESE [16, 17], show that 2 to 6% of the local population is affected. In addition, several studies indicate a link between living near landfills and incinerators and the impoverishment of small areas. However, in the absence of accurate information on the geographic location of landfills, data on social inequalities in exposure related to residence near landfills are less reliable than data for incinerators. This data, along with waste transport activities, will allow an assessment of the overall health impacts of the facilities.

Because the health consequences are nonspecific, both those who live near waste disposal facilities and those who live far from such a facility have essentially the same health problems [18]. The reactions that occur depend on pollutant exposure and personal sensitivity.

Both stronger and more sensitive forms are found in human groups, such as adolescents, pregnant women, and the elderly. Because each person is unique, there is great variation in how resistant or sensitive they are to certain physical and chemical stimuli. Some people living near a landfill might be exposed differently than others, while others might not be affected. The problem is further complicated because it is impossible to determine which cases are specifically related to pollution, even if there is evidence of an increased incidence of adverse health effects near a landfill.

Linking risk to health effects and demonstrating that hazardous substances are escaping from the waste facility and causing health problems for people who live or work nearby are the only ways to find a solution to this complex challenge.

People can meet contaminants from waste facilities in a variety of ways, including inhalation, ingestion of food or water, skin contact, fire, or explosions [3]. Therefore, a hierarchy of exposure data must be established that ranks the exposure assessment in terms of its relationship to actual exposure from best to worst to determine a person's actual exposure.

Existing studies examining the relationship between waste management techniques and health outcomes rely on irrelevant cues, such as employment or living near the site [19, 20]. The research must be based on quantified environmental measurements or individual measurements at the time of likely exposure to maximize the relevance of the evidence. The problem is that most research assumes that the waste facility (incinerators, landfills, and composting) emits pollutants but does not make actual measurements that can be used to determine exposure levels. Therefore, it is extremely difficult to demonstrate a cause-and-effect relationship in epidemiological research with populations exposed to contamination from waste management.

MUNICIPAL SOLID WASTE – LANDFILLING

This paragraph addresses landfilling as the world's primary option for MSW management and the extent to which residents living near landfills may pose a health risk. Risk assessment occurs through a variety of channels, including inhalation of compounds released from the facility, contact with contaminated water or soil, direct ingestion of contaminated products or water, and fire or explosion.

Anyone living near such a landfill is at risk of developing health problems through direct contact with the facility's contaminants by a variety of means, including inhalation, contact with contaminated water or soil, direct consumption of contaminated goods or water, and fire or explosion.

Numerous studies suggest an association between living near or in contact with landfills and health problems [21, 22]. The World Health Organization has published the results of some good studies [21], and the most important finding is that the evidence for an association between landfills and health endpoints (especially cancer, reproductive behavior, and mortality) is either insufficient or inadequate. Although at first glance, one is tempted to assert that the population living near landfills is affected in one way or another. The biggest problems, therefore, should be with illegal, unregulated landfills that accept waste without any sorting at the source.

Exposure Assessment

The quality of exposure assessment may be a reason for insufficient observations, even though studies have shown that there is no clear relationship between landfilling and exposure of the surrounding population to various health risks.

In all available data, distance to the landfill is used as a proxy for exposure; this method can reflect and integrate multiple exposure pathways, including contamination of nearby soil or groundwater, in addition to exposure *via* air. This not only simplifies the calculation, but also retains distance as a common unit of measurement. However, distance is only a proxy for actual exposure to contaminants released from landfills on a first-order basis.

Recent research in this area has proposed to use a special mathematical modeling solution called LandGEM together with a Lagrangian dispersion model to combine geographic distance and predicted concentration emissions [23]. This innovative and intriguing method can simplify things while painting a much clearer picture of how landfills affect the environment and public health. Underlying this method is the assumption that hydrogen sulfide (H_2S) emissions from waste degradation can serve as a signal of air pollution, while removal can serve as an indicator of water and soil matrix pollution.

Health Impacts

Potential health effects of landfills include nuisance, respiratory, cancer and birth defects, and irritation. The most recent studies are from Italy and England and address non-Hodgkin's lymphoma, pancreatic cancer, laryngeal cancer, liver cancer, and kidney cancer, although the totality of evidence is insufficient to make sound judgments [24, 25]. Looking only at municipal solid waste, some negative consequences have been highlighted, but this is less obvious for toxic waste and overall. Numerous studies conducted in the United Kingdom found weak evidence of an association between landfills and increased risk for babies born to mothers living near landfills [25]. Statistical information from the United Kingdom indicates a significantly increased risk of all congenital malformations, neural tube defects, abdominal wall defects, surgical correction of gastroschisis and exomphalos, and low and very low birth weight pregnancies in individuals living within two kilometers of sites that produce both hazardous and non-hazardous waste.

The most recent literature data analyzes health outcomes that are less severe but have a greater overall impact because they occur more frequently in the exposed population [26]. For example, associations have been found between exposure to odor-intensive disposal facilities such as landfills and respiratory symptoms, as well as other nonspecific symptoms such as noise and other nuisance-related problems. Little information is available on the deterioration of mental and social health, such as changes in daily routines or depressive states.

Despite all the uncertainties, the information available today can provide a framework for evaluating the health effects of waste disposal facilities in residential areas. In a health impact assessment (HIA) of municipal solid waste management, health impacts can be selected using a two-step process. First, consideration is given to diseases for which there is "little evidence," such as cancer for landfills, congenital disorders, and low birth weight for incinerators. Because of the lower emissions from incineration, a temporal correction coefficient must be used. Second, consider preterm births at landfills and respiratory illness and nuisance at incinerators based on recent results of multisite studies with at least one similar positive result in the literature [17].

The results that can be considered in the evaluation of landfills are shown in Table **7.1**.

Table 7.1. Measures of exposure and health effects used in evaluating the health effects of landfills.

Exposure Measuring	Exposure Index	Health Outcome	Health Risk	Metrics*
2 km	Distance	congenital anomalies	Relative risk (RR) = 1.02	I.C.
		annoyance from odor	5.4%**	P.
		low birth weight	RR = 1.06	I.C.
5 km	H₂S (disp.model)	respiratory diseases	RR = 1.09	P.

* I.C. = cumulative incidence on the simulation period; P. = annual prevalence
** Confidence intervals, value refers to data from questionnaires.

The formula used for attributable cases (AC) that can be applied is:

$$AC = AF_{exp} \times Rate_{genpop} \times Pop_{exp} \qquad (7.1)$$

Where, $AF_{exp} = \dfrac{RR-1}{RR}$ is the attributable fraction in exposed people, $Rate_{genpop}$ is the background population incidence rate and Pop_{exp} is the exposed population.

Furthermore, AC can be converted to Disability Adjusted Life Years (DALYs), using the formula:

$$DAILY = AC \times DW \times L \qquad (7.2)$$

Where, AC is attributable cases, DW is the disability weight and L is the disease duration.

The methodology is suitable for a first-order approach to estimate the magnitude of health impacts under different waste management scenarios based on the underlying assumptions.

MUNICIPAL SOLID WASTE – INCINERATORS

This paragraph evaluates the health risks associated with the incineration process in relation to the emission of a class of persistent organic chemicals known as "dioxins" Specifically, PCDDs, PCDFs, and PCBs (polychlorinated dibenzo-p-dioxins, polychlorinated dibenzofurans, and polychlorinated biphenyls) (PCBs). PCDDs and PCDFs are formed during incineration processes, especially inadequate incineration of municipal waste.

These compounds accumulate in food (dairy products, eggs, fish and animal fats), are non-biodegradable and quite resistant, and many of them are toxic.

As emissions from incinerators have improved over time, mainly due to tighter restrictions and better emission treatment techniques (so-called best available techniques (BAT)), incineration is now the second most common waste management method in the world after landfilling.

Exposure Assessment

In recent years, there has been a steady decline in incineration emissions, with encouraging data on the decline in anticipated human health hazards. The ratio of concentrations of released substances in the late 2000s compared to the 1990s ranges from one order of magnitude for total suspended solids or some metals *(e.g.,* mercury and cadmium) to four orders of magnitude for dioxins (polychlorinated dibenzo-p-dioxins ($PCDD_s$) and polychlorinated dibenzofurans ($PCDF_s$)). There are several studies that provide analysis comparing emissions over time [27].

Units of toxicity equivalence factor (TEF) are used to measure the toxicity of dioxins. A value of one is assigned for the most hazardous substances, and a value of less than one for all others. Since the effects of the various dioxins are cumulative, the TEF value of each dioxin is multiplied by its concentration to obtain a final toxicity equivalent quantity.

The most recent exposure assessment techniques used in epidemiologic studies of incinerators categorize the effectiveness of exposure methods according to three factors: the proxy measure chosen for exposure intensity (qualitative or quantitative measure or model), the scale at which the spatial distribution of the exposed population was considered (community, small area, residential address), and whether temporal variability in exposure was considered. When all three aspects were considered, the exposure assessment methods improved, with a reduction in the misclassification of exposure [28].

Recent studies have highlighted the use of human biomonitoring as a more accurate exposure indicator than environmental monitoring, *e.g.,* levels of polycyclic aromatic hydrocarbons (PAH) in the urine associated with increasing emission levels from incinerators. Compared to other studies that found no evidence of adverse health effects, human biomonitoring studies show higher dioxin levels in people living near incinerators [29]. These results support the use of human biomonitoring in waste incineration to improve the assessment of exposure to low-level environmental stressors.

Health Impacts

Because the evidence is limited to the study period and the various incinerators studied, it is more difficult to make general considerations about health effects. This is primarily because incinerator emissions have changed over time, which affects health effects (old generation versus new generation of facilities). Developing exposure assessment procedures can help summarize the health concerns associated with waste incineration in general.

Based on quantitative estimates of additional risks, past studies of the health effects of incinerators have indicated a measurable risk of various cancers (gastric, colon, liver, and lung) to the population living nearby [28]. The main problem with these studies is that they used data from old generation incinerators that had high emission levels. Emissions from today's incinerators differ in quantity and composition due to the availability of modern treatment technologies. For this reason, it is impossible to analyze and use the results of all available studies, which makes it extremely difficult to assess the impact of waste incineration on human health.

According to some data, increasing exposure to incinerators is associated with poorer pregnancy outcomes, such as preterm birth and spontaneous abortion [30, 31].

EPIDEMIOLOGICAL INVESTIGATIONS

Epidemiological studies of the effects of waste management practices on human health are, for ethical reasons, observational studies only and not experimental studies. Experimental studies are some clinical trials in which a test population is exposed to a substance or drug and a control population is not, and the expected outcome is usually positive [32].

The most used types of investigations can be summarized as:

o In *prospective cohort studies*, there are two cohorts of participants: those who were exposed and those who were not. These studies extend over an extended period, tracking the extent of population exposure and the rate of disease development. If necessary, other data is also collected through surveys. The primary source of data collected and studied is human fluids or tissues. The basic disadvantage of these studies is that a large population must be collected to control possible confounding factors and to ensure the statistical significance of the results, which usually results in high costs.

o *Retrospective case-control studies*, in which patients who have already developed a particular disease are selected as the case group and a control group of healthy individuals. These studies are usually less expensive than prospective cohort studies because they involve smaller participant groups, require fewer investigators, and usually rely on interviews to obtain data on past exposure. Because fewer individuals participate in the study, the representativeness of the results may be questioned, which is a weakness of these studies.

o *Cross-sectional studies* are the counterpart of longitudinal studies that examine a specific portion of the exposed population over a short period of time. These studies can be helpful in generating hypotheses that can later be tested by more in-depth studies. The disadvantage of these studies is that they are usually less costly and may prove beneficial if the disease being studied is relatively common. Because it can be difficult to determine whether a particular condition occurred before or after the group was exposed to a potential hazard, many factors are considered in these types of studies.

In the event of a major accident, human exposure to substances released from a waste facility may be acute, resulting in short-term exposure to high concentrations of potentially hazardous substances, ionizing radiation, bioaerosols, or dust, or chronic, resulting in prolonged exposure to low concentrations of these substances or radiation. The task of epidemiologists becomes even more difficult at sites with landfills, incinerators, or other waste management facilities that are state of the art, constructed with the best available technology, and operated in compliance with regulations and in full regulatory compliance. Therefore, the study must have sufficient statistical data to avoid erroneous positive or negative conclusions and can detect significant clinical differences between a control group and a "test" population. The validity of the study is highly dependent on the sample size, even though the incidence of certain clinical consequences does not usually differ between the two populations. Consequently, studies must be conducted on at least tens of thousands of individuals in both the exposed area and the control area. The resources required for such studies are rarely accessible, and the theoretical population size required for statistically correct interpretation of the data may be larger than the entire population in the area under study. A comprehensive analysis incorporating the results of numerous individual studies may be a compromise solution despite its own shortcomings *(e.g.,* the difficulty of controlling for bias in the original studies, the difficulty of accessing studies that did not show statistically significant results and therefore remained unpublished).

Epidemiological studies, as mentioned, have many limitations, in addition, they also include:

• Insufficient information on emissions.

• No data available on the effects of direct exposure to pollution from landfills and other sources.

• Confounding factors *(e.g.,* ethnicity, income, exposure from other sources).

• Movement of the population.

• Some diseases have a long latency period.

To overcome these limitations, several practical recommendations and guidelines have been developed to help prevent the dissemination of biased information [33, 34].

By calculating the ratio between the incidence of disease in the exposed population and the incidence of the same disease in the unexposed population, epidemiologists can determine the strength of the association between exposure to a potentially harmful substance and certain health effects. This is referred to as relative risk (RR), whereas odd risk is the measure of association used in case-control research (OR).

For a man at higher risk for a particular health outcome, the RR value would be above 1. For example, if the RR is 5, the risk is 400% (or five times) higher. Even though a particular exposure condition carries a higher risk (RR > 1) for a particular disease, the exact reason for the health effects may not yet be known.

To determine whether a pollutant has an adverse effect on health, many additional factors must be considered. One of these is the statistical significance of the discovered association, since one must attempt to rule out the possibility that the association is the result of chance. The confidence level for this strategy is between 95% and 99% in most cases. The approach used by the World Cancer Research Fund and the American Institute for Cancer Research [35] to determine the level of evidence supporting an association between exposure and disease is shown in Table **7.2**.

The combined evaluation of the RR and statistical significance found yields the risk level values. Although most epidemiological studies of health effects possibly related to waste management report RR or OR values of less than 1.5 and rarely more than 2, high RR values indicate considerable evidence of adverse health effects associated with certain environmental factors [36].

Table 7.2. The relative risk (RR) and odd ratio (OR) model.

RR or OR	Statistical Significance	Strength of Evidence
0.87–1.5	NO	No association
1.5–2.0	NO	No association
1.5–2.0	YES	Moderate
>2	NO	Moderate
>2	YES	Strong

WASTE COMPOSTING

This paragraph presents the main concern about composting in terms of health impacts associated with respiratory and dermal illnesses. Respiratory illnesses can be caused by exposure to dust and bacteria, fungi, and actinomycetes. In addition, epidemiological studies show an association between irritant respiratory symptoms in residents and bioaerosol exposure from outdoor composting facilities.

Most of the available information indicates that the greatest concern is for compost workers, as they are more likely than the general population to suffer from respiratory and skin diseases. Outdoor windrow systems are used to process all recyclable household waste and green waste that can be composted. Exposure to dust, bacteria, fungi, actinomycetes, endotoxins, and 1-3 glucans released during composting can lead to respiratory diseases [37].

Little research has been done on the effects of composting facilities on the health of the surrounding population, especially regarding the dispersal of bioaerosols. Composting has several adverse health effects, including upper respiratory tract inflammatory reactions *(e.g.,* runny nose, sore throat, and dry cough), toxicoses *(e.g.,* toxic pneumonitis caused by endotoxins, respiratory infections, and skin infections), and allergies *(e.g.,* bronchial asthma, allergic rhinitis, extrinsic allergic alveolitis) [38].

Due to our lack of understanding of dose-response relationships and the risks associated with bioaerosols, there is no definition of a safe buffer distance. According to certain studies, there is a relationship between the irritable respiratory symptoms of residents and bioaerosol exposure from outdoor composting facilities [39].

DISCUSSION POINTS

The generation, management, and disposal of waste involve more complicated activities that can directly and indirectly harm health through various pathways and mechanisms. In addition to affecting well-being through odor nuisance, health effects include increased risk of cancer and mortality, respiratory disease, congenital malformations, and low birth weight.

Further studies are needed to improve the evidence base needed to develop solutions, as the causal relationship between waste management practices and health outcomes is only partially understood, and data are generally inconclusive. The summary of waste management activities and health impacts is presented in Table **7.3**.

Table 7.3. Summary of the waste management activity and health outcomes through exposure route and hazard.

Waste Management Activity	Exposure Route	Hazard	Health Outcome
Composting	Exposure to a central composting facility at work, proximity to a central composting facility at home	Bioaerosols with bacteria *(e.g.,* Clostridium botulinum, endotoxin-producing gram-negative bacteria) and/or fungal spores (Aspergillus fumigatus)	Airways symptoms
Landfilling	Residence near site, occupational exposure	Any hazards – organic compounds, heavy metals, *etc.*	Any health outcomes
Incineration	Residence near site, occupational exposure	Any hazards - heavy metals, organic compounds such as dioxins, *etc.*	Any health outcomes

The purpose of many epidemiological studies is not to prove causality but to hypothesize and point out potential problems. Given their great influence on policy discussions, more thorough studies of little-studied but potentially important health outcomes such as neurological disease and odor nuisance are sometimes warranted. For more informative and complete epidemiological studies, a better methodology for exposure assessment is needed. Human biomonitoring for persistent substance identification can represent a powerful tool in this direction.

In addition to the limited knowledge, the available evidence is becoming less available and relevant for some countries, as the waste industry evolves. The available data on the health impacts of different types of waste management are from old facilities, especially landfills and incinerators. With the advancement of modern technology, pollutant emissions have decreased significantly, and their detectable health effects have often diminished. Overall, the best method for elucidating these historical trends would be a longitudinal analysis, ideally on population cohorts.

In many countries, old generation facilities are still in operation, and health impacts can probably be compared with those described in the literature. In addition, because of the nature of the waste flows that are changing, the impact can be bigger.

Sustainability in various sectors, including waste management, is one of the agenda items of the Sustainable Development Goals (SDGs) [40]. Although there is a lack of convincing information about specific health impacts, sustainability has more complex components and can still serve as a guiding concept for policy development. The Sustainable Development Goals (SDGs) can be useful in real-world situations, such as when local governments must decide on regulations for waste management.

To fill existing information gaps in this area and prevent new ones, methods and resources for effective health surveillance should be established.

CONCLUDING REMARKS

Direct and indirect health implications of waste technologies include odor annoyance, cancer risks, respiratory illnesses, and birth difficulties. However, because of conflicting evidence, our knowledge of the waste-health link is insufficient. Conseguently, more investigation is required, concentrating on unstudied effects such as neurological disorders and odor. Studies can be improved by using more advanced exposure assessment techniques, such as biomonitoring. While modern technology has decreased pollutant emissions and associated discernible health implications, the majority of the research now available originates from out-of-date waste facilities. Historical patterns can be found through longitudinal analysis of population cohorts. Sustainability is important, but specific health effects demand further research. Although the Sustainable Development Goals (SDGs) serve as a framework for policy creation, thorough health surveillance is essential to closing knowledge gaps and promoting efficient waste management for public health.

REFERENCES

[1]   "WHO Regional Office for Europe Population health and waste management: Scientific data and policy options.Copenhagen: WHO Regional Office for Europe", Available at: http://www.euro.who.int/__data/assets/pdf_file/0012/91101/E91021.pdf.

[2]   A.K. Ziraba, T.N. Haregu, and B. Mberu, "A review and framework for understanding the potential impact of poor solid waste management on health in developing countries", *Arch. Public. Health.,* vol. 74, no. 1, p. 55, 2016.

[http://dx.doi.org/10.1186/s13690-016-0166-4]

[3]   D. Porta, S. Milani, A.I. Lazzarino, C.A. Perucci, and F. Forastiere, "Systematic review of epidemiological studies on health effects associated with management of solid waste", *Environ. Health.,* vol. 8, no. 1, p. 60, 2009.
[http://dx.doi.org/10.1186/1476-069X-8-60]

[4]   P. Xu, Z. Chen, L. Wu, Y. Chen, D. Xu, H. Shen, J. Han, X. Wang, and X. Lou, "Health risk of childhood exposure to PCDD/Fs emitted from a municipal waste incinerator in Zhejiang, China", *Sci. Total. Environ.,* vol. 689, pp. 937-944, 2019.
[http://dx.doi.org/10.1016/j.scitotenv.2019.06.425]

[5]   M. Vaccari, G. Vinti, and T. Tudor, "An analysis of the risk posed by leachate from dumpsites in developing countries", *Environments,* vol. 5, no. 9, p. 99, 2018.
[http://dx.doi.org/10.3390/environments5090099]

[6]   P. Negi, S. Mor, and K. Ravindra, "Impact of landfill leachate on the groundwater quality in three cities of north india and health risk assessment", *Environ. Dev. Sustain.,* vol. 22, no. 2, pp. 1455-1474, 2020.
[http://dx.doi.org/10.1007/s10668-018-0257-1]

[7]   W.B. Kindzierski, and S. Gabos, "Health effects associated with wastewater treatment, disposal and reuse", *Water Environ. Res.,* vol. 68, no. 4, pp. 818-826, 1996.
[http://dx.doi.org/10.2175/106143096X135687]

[8]   S.W. Hu, and C.M. Shy, "Health effects of waste incineration: A review of epidemiologic studies", *J. Air Waste Manag. Assoc.,* vol. 51, no. 7, pp. 1100-1109, 2001.
[http://dx.doi.org/10.1080/10473289.2001.10464324]

[9]   "Environment agency", In: *Environment Agency Health Effects of Composting : A Study of Three Composting Sites and Review of Past Data.* AEAT: London, 2001.

[10] "Directive 2008/98/EC of the European Parliament and of the Council of 19 November 2008 on Waste and Repealing Certain Directives. Official Journal of the European Union; L(312):3 30", Available at: http://eur-lex.europa.eu/legal-content/EN/TXT/?uri=CELEX:32008L0098.

[11] P.W. Tait, J. Brew, A. Che, A. Costanzo, A. Danyluk, M. Davis, A. Khalaf, K. McMahon, A. Watson, K. Rowcliff, and D. Bowles, "The health impacts of waste incineration: A systematic review", *Aust. N. Z. J. Public Health,* vol. 44, no. 1, pp. 40-48, 2020.
[http://dx.doi.org/10.1111/1753-6405.12939]

[12] A. Perteghella, G. Gilioli, T. Tudor, and M. Vaccari, "Utilizing an integrated assessment scheme for sustainable waste management in low and middle-income countries: Case studies from Bosnia-Herzegovina and Mozambique", *Waste Manag.,* vol. 113, pp. 176-185, 2020.
[http://dx.doi.org/10.1016/j.wasman.2020.05.051]

[13] A. Redfearn, and D. Roberts, "Health effects and landfill sites", In: *Environmental and Health Impact of Solid Waste Management Activities. Issues in Environmental Science and Technology.,* R.E. Hester, R.M. Harrison, Eds., vol. Vol. 18. Royal Society of Chemistry: Cambridge, UK, 2002, pp. 103-140.

[15] S. Kaza, L.C. Yao, P. Bhada-Tata, and F. Van Woerden, *What a Waste 2.0: A Global Snapshot of Solid Waste Management to 2050; Urban Development.* World Bank: Washington, DC, USA, 2018.
[http://dx.doi.org/10.1596/978-1-4648-1329-0]

[16] A. Ranzi, C. Ancona, P. Angelini, C. Badaloni, A. Cernigliaro, and M. Chiusolo, "Health impact assessment of policies for municipal solid waste management: Findings of the SESPIR Project", *Epidemiol. Prev.,* vol. 38, no. 5, pp. 313-322, 2014.

[17] F. Forastiere, C. Badaloni, K. de Hoogh, M.K. von Kraus, M. Martuzzi, F. Mitis, L. Palkovicova, D. Porta, P. Preiss, A. Ranzi, C.A. Perucci, and D. Briggs, "Health impact assessment of waste management facilities in three European countries", *Environ. Health,* vol. 10, no. 1, p. 53, 2011.
[http://dx.doi.org/10.1186/1476-069X-10-53]

[18]   L. Giusti, "A review of waste management practices and their impact on human health", *Waste Manag.,* vol. 29, no. 8, pp. 2227-2239, 2009.
        [http://dx.doi.org/10.1016/j.wasman.2009.03.028]

[19]   B. Parkes, A.L. Hansell, R.E. Ghosh, P. Douglas, D. Fecht, D. Wellesley, J.J. Kurinczuk, J. Rankin, K. de Hoogh, G.W. Fuller, P. Elliott, and M.B. Toledano, "Risk of congenital anomalies near municipal waste incinerators in england and scotland: Retrospective population-based cohort study", *Environ. Int.,* vol. 134, p. 104845, 2020.
        [http://dx.doi.org/10.1016/j.envint.2019.05.039]

[20]   F. Mataloni, C. Badaloni, M.N. Golini, A. Bolignano, S. Bucci, R. Sozzi, F. Forastiere, M. Davoli, and C. Ancona, "Morbidity and mortality of people who live close to municipal waste landfills: A multisite cohort study", *Int. J. Epidemiol.,* vol. 45, no. 3, pp. 806-815, 2016.
        [http://dx.doi.org/10.1093/ije/dyw052]

[21]   "WHO, 2007.Population health and waste management: scientific data and policy options. Report of a WHO Workshop, Rome, Italy, 29–30 March 2007", *World Health Organisation (WHO), European Centre for Environment and Health.*

[22]   P. Elliott, S. Richardson, J.J. Abellan, A. Thomson, C. de Hoogh, L. Jarup, and D.J. Briggs, "Geographic density of landfill sites and risk of congenital anomalies in England", *Occup. Environ. Med.,* vol. 66, no. 2, pp. 81-89, 2008.
        [http://dx.doi.org/10.1136/oem.2007.038497]

[23]   US EPA, "Landfill gas emissions model (landgem) version 3.02 user's guide. research triangle park, NC: United states environmental protection agency", Available at: https://www3.epa.gov/ttncatc1/dir1/landgem-v302-guide.pdf

[24]   O. Paladino, and M. Massabò, "Health risk assessment as an approach to manage an old landfill and to propose integrated solid waste treatment: A case study in Italy", *Waste Manag.,* vol. 68, pp. 344-354, 2017.
        [http://dx.doi.org/10.1016/j.wasman.2017.07.021]

[25]   P. Elliott, S. Richardson, J.J. Abellan, A. Thomson, C. de Hoogh, L. Jarup, and D.J. Briggs, "Geographic density of landfill sites and risk of congenital anomalies in England", *Occup. Environ. Med.,* vol. 66, no. 2, pp. 81-89, 2008.
        [http://dx.doi.org/10.1136/oem.2007.038497]

[26]   G. Vinti, V. Bauza, T. Clasen, K. Medlicott, T. Tudor, C. Zurbrügg, and M. Vaccari, "Municipal solid waste management and adverse health outcomes: A systematic review", *Int. J. Environ. Res. Public Health,* vol. 18, no. 8, p. 4331, 2021.
        [http://dx.doi.org/10.3390/ijerph18084331]

[27]   M. Vinceti, C. Malagoli, S. Fabbi, S. Teggi, R. Rodolfi, L. Garavelli, G. Astolfi, and F. Rivieri, "Risk of congenital anomalies around a municipal solid waste incinerator: A GIS-based case-control study", *Int. J. Health Geogr.,* vol. 8, no. 1, p. 8, 2009.
        [http://dx.doi.org/10.1186/1476-072X-8-8]

[28]   S. Candela, A. Ranzi, L. Bonvicini, F. Baldacchini, P. Marzaroli, A. Evangelista, F. Luberto, E. Carretta, P. Angelini, A.F. Sterrantino, S. Broccoli, M. Cordioli, C. Ancona, and F. Forastiere, "Air pollution from incinerators and reproductive outcomes: A multisite study", *Epidemiology,* vol. 24, no. 6, pp. 863-870, 2013.
        [http://dx.doi.org/10.1097/EDE.0b013c3182a712f1]

[29]   N. Linzalone, and F. Bianchi, "Human biomonitoring to define occupational exposure and health risks in waste incinerator plants", *Int. J. Environ. Health,* vol. 3, no. 1, pp. 87-105, 2009.
        [http://dx.doi.org/10.1504/IJENVH.2009.022907]

[30]   S.W. Hu, and C.M. Shy, "Health effects of waste incineration: A review of epidemiologic studies", *J. Air Waste Manag. Assoc.,* vol. 51, no. 7, pp. 1100-1109, 2001.
        [http://dx.doi.org/10.1080/10473289.2001.10464324]

[31]   D.C. Ashworth, P. Elliott, and M.B. Toledano, "Waste incineration and adverse birth and neonatal outcomes: A systematic review", *Environ. Int.,* vol. 69, no. August, pp. 120-132, 2014.
[http://dx.doi.org/10.1016/j.envint.2014.04.003]

[32]   L. Rushton, and P. Elliott, "Evaluating evidence on environmental health risks", *Br. Med. Bull.,* vol. 68, no. 1, pp. 113-128, 2003.
[http://dx.doi.org/10.1093/bmb/ldg020]

[33]   K.F. Schulz, and D.A. Grimes, "Sample size calculations in randomised trials: Mandatory and mystical", *Lancet,* vol. 365, no. 9467, pp. 1348-1353, 2005.
[http://dx.doi.org/10.1016/S0140-6736(05)61034-3]

[34]   D. Moher, K.F. Schulz, and D.G. Altman, "The CONSORT statement: Revised recommendations for improving the quality of reports of parallel-group randomised trials", *Lancet,* vol. 357, no. 9263, pp. 1191-1194, 2001.
[http://dx.doi.org/10.1016/S0140-6736(00)04337-3]

[35]   "Nutrition and the prevention of cancer: A global perspective", In: *World Cancer Research Fund (WCRF)* American Institute for Cancer Research (AICR): Washington, DC, USA, 1997.

[36]   L. Tomatis, *Cancer: Causes, Occurrence and Control* International Agency for Research on Cancer: France, 1990.

[37]   Environment Agency, *Environment Agency Health Effects of Composting : A Study of Three Composting Sites and Review of Past Data.* AEAT: London, 2001.

[38]   C.S. Clark, H.S. Bjornson, J. Schwartz-Fulton, J.W. Holland, and P.S. Gartside, "Biological health risks associated with the composting of wastewater treatment plant sludge", *J. Water Pollut. Control Fed.,* vol. 56, no. 12, pp. 1269-1276, 1984.

[39]   M.C. Maritato, E.R. Algeo, and R.E. Keenan, "Potential human health concerns from composting", *Biocycle,* vol. 33, no. 12, p. 70, 1992.

[40]   L. Rodić, and D. Wilson, "Resolving governance issues to achieve priority sustainable development goals related to solid waste management in developing countries", *Sustainability,* vol. 9, no. 3, p. 404, 2017.
[http://dx.doi.org/10.3390/su9030404]

CHAPTER 8

Concluding Remarks

Diana Mariana Cocârţă[1,*]

[1] *Department of Energy Production and Use, Faculty of Energy Engineering, University POLITEHNICA of Bucharest, Bucharest, Romania*

Environmental pollution has been a topic of growing interest all over the world in both developed and developing countries. Fast demographic increase in different regions, demand for energy, food production, machine development, and increased trend of urbanization have resulted in serious pollution of soil, water, and air that affects the surrounding environment and human health. According to the World Health Organization (WHO), 24% of all estimated global mortalities are linked to environmental pollution. On the other hand, environmental pollution has led to serious disruptions in natural systems, *e.g.,* snow and ice are melting, hydrological and biological systems are changing, and negative pollution effects are not stopping here.

The main objective of the proposed book is to enlarge the understanding of the three major environmental systems (soil, water, and air) from an environmental issues perspective, and their management. The chapters are centered on a risk-based approach to environmental issues and go in-depth into risk management implementation, human and ecological risk assessment of contaminated sites, risk-based approach for the management of contaminated sites, air quality and assessing the risks from air pollution, risk assessment for protection of drinking water quality, and the importance of risk analysis in waste management. They also provided some good practices while considering environmental risks and instruments for assessing risks to human health and the environment.

The presented theoretical information covers specific terminology on Environmental Pollution Health Impact and Health Risk Assessment.

Within the book content, it is illustrated how to apply the assessment process and how to evaluate the risks at different scales, regardless of whether it is about air pollution, the management of contaminated sites, and the identification of the remedial strategy, safe drinking water, or waste management. The results

---

[*] **Corresponding author Diana Mariana Cocârţă:** University POLITEHNICA of Bucharest, Faculty of Energy Engineering, Splaiul Independentei 313, RO-060042 Bucharest, Romania; E-mail: dianacocarta13@yahoo.com

provided by the health risk assessment tools can be transposed into local, regional, or national policies aiming to reduce the air pollution impact on human health.

The current book enables aspiring environmental protection specialists (students) to understand the systems and procedures for assessing the harmful effects of environmental pollution caused by human activities on population health and the environment. This manuscript offers a comprehensive set of procedures to analyze the negative effects of environmental problems by in-field professionals who are actively involved in environmental pollution control. Decision-makers in both public and private sectors should find the volume useful. Implementing the knowledge from the book makes it easier to create environments that are sustainable and healthy.

ACKNOWLEDGEMENTS

This work of developing the eBook content was supported by the Erasmus+ Programme SafeEngine project, contract no 2020-1-RO01-KA203-080085. The European Commission's support for this publication does not constitute an endorsement of the contents, which reflects the views of the authors, and the National Agency and Commission cannot be held responsible for any use which may be made of the information contained therein.

SUBJECT INDEX

A

Acid
  ethylenediaminetetraacetic 74
  hydrochloric 74
  nitrilotriacetic 74
Actinomycetes 152
Activities
  anthropogenic 10
  economic-industrial 46
  metabolic 78
  waste transport 144
Acute arsenic poisoning 50
Air pollutant(s) 16, 92, 93, 100
  dangerous 92
  exposures 102
  toxic 102
Air pollution 1, 3, 5, 15, 87, 88, 89, 90, 101,
    102, 103, 107, 108, 141, 158
  health risk assessment 90, 102
Air quality 3, 4, 5, 6, 9, 87, 98, 102, 158
  management plans 97
  measurements 103
Allergic rhinitis 152
Alzheimer's disease 49
Anthropic sources 19
Anthropogenic sources 10
Arsenicosis 50
Arsenic toxicity 50
Aspergillus fumigatus 153
Asthma, bronchial 152

B

Bradykinesia 52
Breast cancer risk and chronic disease 53

C

Cancer
  liver 146

lung 15, 23, 88
  pancreatic 146
Cardiovascular
  diseases 15, 16, 19, 92
  effects 3
Chelating agents 77
Chemical(s) 9, 10, 12, 29, 32, 41, 45, 61, 63,
    92, 96, 113
  contaminants 1, 94
  toxicity information 29
Chronic
  daily intake (CDI) 31
  diseases 53, 96
  obstructive pulmonary disease 3
  pulmonary diseases 15
Clostridium botulinum 153
Cognitive
  dysfunction 52
  function 52
Conditions 12, 17, 37, 43, 50, 57, 59, 63, 68,
    88
  anaerobic 68
  meteorological 11
  neurological 52
  thermophilic 78
Contaminants
  inorganic 74, 112
  organic 112
Contaminated soil 61, 65, 67, 70, 74, 78
Contamination 11, 16, 112, 114, 117, 118,
    124, 129, 131, 144, 145
  chemical 113
  heavy metal 63
  microbial 113
  radioactive 55
  radiological 55

D

Damage 14, 37, 40, 42, 43, 51, 136
  environmental 23, 56
  kidney 50

www.ingramcontent.com/pod-product-compliance
Lightning Source LLC
Chambersburg PA
CBHW041706210326
41598CB00007B/550